Cuthbert Collingwood

Rambles of a Naturalist on the Shores and Waters of the China Sea

Being observations in natural history during a voyage to China, Formosa, Borneo, Singapore, etc., made in Her Majesty's vessels in 1866 and 1867

Cuthbert Collingwood

Rambles of a Naturalist on the Shores and Waters of the China Sea
Being observations in natural history during a voyage to China, Formosa, Borneo, Singapore, etc., made in Her Majesty's vessels in 1866 and 1867

ISBN/EAN: 9783337243593

Printed in Europe, USA, Canada, Australia, Japan

Cover: Foto ©berggeist007 / pixelio.de

More available books at **www.hansebooks.com**

Sandstone Pillars, South Side of Kelung Harbour, Formosa.

Frontispiece.

RAMBLES OF A NATURALIST

ON

THE SHORES AND WATERS OF THE CHINA SEA:

BEING

OBSERVATIONS IN NATURAL HISTORY DURING A VOYAGE
TO CHINA, FORMOSA, BORNEO, SINGAPORE, Etc.,
MADE IN HER MAJESTY'S VESSELS
IN 1866 AND 1867.

BY

CUTHBERT COLLINGWOOD, M.A., M.B., Oxon.,
F.L.S., *ETC.*

LONDON:
JOHN MURRAY, ALBEMARLE STREET.
1868.

Dedication.

TO

JOSEPH DALTON HOOKER, D.C.L., M.D., F.R.S., &c.

DIRECTOR OF THE ROYAL GARDENS, KEW.

DEAR DR. HOOKER:

I know that it is to you that I am mainly indebted for the opportunity of making the observations contained in the following pages, as well as for the pleasure I have derived from the increased scope of my Natural History studies which my voyage has afforded me. And although the tenor of those observations and studies has been in the main zoological rather than botanical, I will not permit that circumstance to deprive me of the gratification of dedicating the results to one who so kindly and readily aided me with his influence and advice.

I am, dear Dr. Hooker,

Sincerely yours,

CUTHBERT COLLINGWOOD.

GREENWICH,
March, 1868.

PREFACE.

THE circumstances under which the voyage here recorded was undertaken need not be fully entered into. Suffice it to say, that, actuated solely by a desire of increasing my own information, and the hope of, in some measure, advancing science, I was induced to seize an opportunity which seemed to present itself of fulfilling what had always been an object of my ambition. The pleasure I have myself derived from it has entirely obscured the vexations and drawbacks to which I have been subjected in its fulfilment. That I have met with disappointment and discouragement from those to whom I had most right to look for support and co-operation, is a circumstance which, although it greatly limited and curtailed my operations, is entirely forgotten in the delight of having visited Nature in her deepest recesses, and viewed her in her grandest aspects.

Whatever slight value the following pages may possess, will be due to the circumstance that the facts they record are derived from observation, and not from books; and that I have not endeavoured to adjust my own observations to the experience of others, but have rather corrected my previous knowledge by the aid of personal research.

I feel bound in this place to tender my thanks to certain gentlemen who have kindly aided me in my plans and movements:—To Captain Richards, Hydrographer of the Navy, to whose recommendation I am indebted for the primary opportunity of making the voyage; to Commander Bullock, of H.M.S. "Serpent," and Captain Courtenay, late of H.M.S. "Scylla," from both of whom I received uniform courtesy and kindness; to Lieutenant Richards, and Mr. Sutton (Chief Engineer) of the "Serpent," and Lieutenant D. Stewart, of the "Scylla," to each of whom I am indebted for steady and valuable assistance; to Mr. Josè d'Almeida, of Singapore, Messrs. Hugh Low, C. C. De Crespigny, J. Tyndall Woods, and Howard, of Labuan; to Mr. Alfred Houghton, Mr. Martin, and the Tuan Muda of Sarawak; to Dr. Maxwell, of Ta-kau, and Mr. Gregory, Vice-Consul of Tam-suy, Formosa; to the Venerable Archdeacon Gray, Consular Chaplain, and Dr. J. G. Kerr, of Canton; and last, though not least, to Mr. F. D. Lalcaca, and (the late) Captain Jameson, of Hong Kong;—all of whom showed me various acts of kindness, which will not soon be forgotten.

My thanks are also due to Drs. Baird and Günther; to Messrs. G. R. Gray, F. Walker, Frederick Smith, A. G. Butler, and Waterhouse, Junior, of the British Museum; as well as to Messrs. Albany Hancock, C. Spence Bate, and Professor Oliver;—all of whom have kindly assisted me in the identification of species.

It should be mentioned, that the account of the Pratas Island, and the chapter on the Luminosity of the Sea, were published in the "Quarterly Journal of Science" for 1867, and are reprinted with the permission of the Editors, having had the advantage of subsequent revision and enlargement. Other papers have been incorporated from the Proceedings of the Linnæan, Geological, Ethnological, and Royal Geographical Societies, as well as from the Annals of Natural History, &c., in order to give unity to the narrative, and completeness to the Natural History observations.

GREENWICH,
March, 1868.

CONTENTS.

CHAPTER I.

THE VOYAGE TO CHINA OVERLAND—IMPRESSIONS OF HONG KONG.

PAGE

Dissolving Views—Marseilles—The Bear Rock—Flocks of Cranes—Alexandria—The Delta—Grand Cairo—The Desert—Red Sea—Aden—First Shore Hunt—Point de Galle—Tropical Calm—Lightning—Tropic Birds—Singapore—Traveller's Tree—Caricature Plant—Approach to Hong Kong—Appearance from the Sea—Boats and their occupants 1

CHAPTER II.

HONG KONG TO PRATAS ISLAND.

Anglified appearance of Hong Kong—Physical Character—The Chinese Quarter—Leave Hong Kong—Arrive at Pratas Reef—Description of the Island—We Visit the Island—Its Vegetation—Insect Fauna—Marine Animals—Sea-Weeds—Fishermen's Temple—Lagoon—Birds of the Island—The Gannets' Settlement—The Seine—Towing Net—Rollers—We quit the Reef—Birds observed upon the Ship between Pratas and Formosa 20

CHAPTER III.

FORMOSA.
TA-KAU-CON, AND THE PESCADORES ISLAND.

Character of Native Race—Dutch Occupation—Treaty Ports—East Coast—Arrive at Ta-kau—Lagoon—Apes' Hill—Land-Crabs—Leaping Fishes—Walk in the Country—Water Buffaloes—Padi Birds—Village of Pi-hi-kun—Chinese Ladies—The Pescadores—

viii   CONTENTS.

                                                                             PAGE

Ponghou—Makung—Cheap Provisions—Cuttle Fish—Absence of Trees and Birds—The Rocks—Visit the Mandarin—Photography—Wreckers 35

CHAPTER IV.

FORMOSA (*continued*)—TAM-SUY.

Towing Net in Formosa Channel—Pterosoma—Firola—Sagitta—Atlanta—Glaucus—Alima—Phyllosoma, or glass-crab—Cerapus—Hyalæa—West Coast of Formosa—Fort Zeelandia—Notonectæ—Arrive at Tam-suy—The Harbour—Boulder Clay—Chinese Graves—Rice-paper Plant—Bamboo—The Town—People—Rice Embargo—Visit to Mbang-ka—Camphor Monopoly—Visit the Chief Mandarin—Return Visit—Queen's Birthday 54

CHAPTER V.

FORMOSA (*continued*)—FROM TAM-SUY TO KE-LUNG.

The Sulphur Springs near Tam-suy—approach to them—their present condition—effects on Animal Life—Preparations for River Voyage—Village of Pah-chie-nah—Arrive at Sik-kow—Bivouac at Chuy-teng-cha—Birds on the Route—Rapids—Population—Domestic Animals—Arrive at Liang-kha—Descent to Ke-lung—Character of the People 70

CHAPTER VI.

FORMOSA (*continued*)—KE-LUNG.

Prevalence of Sandstone—Formation of the Harbour—Caverns—Village Population—Modes of Fishing—Sandstone Peaks and Images—Rising of the Coast—The Coal Mines—Mode of Working—Value of the Coal—Geological position of the Beds—Burning Properties—Petroleum—Marine Animals of the Shore—Peronia—Aplysia—Nudibranchs—Creseis—Singular shoal of Stephanomias . . . 85

CHAPTER VII.

FORMOSA (*continued*)—SAU-O BAY.

East Coast—Steep Island—Reefs at Sau-o—Chinese Village of Sau-o—Village of Tame Aborigines—Their Huts—Physical Characters—Dress—Native Cloth—Search after the Wild Aborigines—Characteristics of the Villagers—Their Occupations—An Alarm—They visit the Ship—Native Politeness—Language—Religious Ideas—Diseases—Distinctions from Chinese Race 101

CHAPTER VIII.

THE ISLANDS NORTH-EAST OF FORMOSA.

Visit of a Chinese Admiral—Ke-lung Island—The Harbour from the Sea—Pinnacle Island—Craig Island—The Wideawakes—Their Breeding Place—Geological Structure of Craig Island—Hunt on the Rocks—Grapsi—Agincourt Island—Pinnacle Rocks—Hoa-pin-san and Tia-usu—The Raleigh Rock—The Dredge—Chromodoris—Gigantic Foraminifera—Further Search—Return to Ke-lung . . 116

CHAPTER IX.

HAITAN STRAITS AND COAST OF CHINA.

Red Discoloration of the Sea—Haitan Island and Straits—Middle Island—New Anemone—Black Islet—Its Fauna and Flora—Chinese Pirates—Rumbling Fish—Slut Island—New Nudibranchs—Iridescent Seaweed—Trigger-Shrimp—Comatula—The River Min—Pagoda Anchorage—Chinese Pagodas—Shwin-gan Passage—Luminous Sea—Plague of Flies—Insects at Sea—Wosung River—Shanghai 129

CHAPTER X.

HONG KONG TO LABUAN.

Atmospheric Phenomena—Fiery-Cross Reef—Corals and Coral Fish—A Wade on the Reef—Marine Animals—Gigantic Anemones—Anemone-inhabiting Fish—Stormy Weather—Waterspout—Aspect of Labuan—Vegetation—The Jungle—Camphor Trees—The Coal Mines—Workings—Quality of the Coal—Geological Considerations—Petroleum 145

CHAPTER XI.

LABUAN.

Bruni, the Capital of Borneo—Piracy—Establishment of the Colony of Labuan—Its Objects—Natural Productions—Pigs—Monkeys—

CONTENTS.

Kahan, or Proboscis Monkey—Birds—Megapode—Chick-chack-Barking Lizard—Iguanas—Cobra—Pythons—Electric Snake—Scorpions—Centipedes—Cicadas—Beetles—Hemiptera—Desecration of European Graves—Isolated Position of the Residents of Labuan 162

CHAPTER XII.

LABUAN (*continued*).

Butterflies of Labuan—Mode of Flight—Number of Species—Dominant Species—Butterflies of Pulo Daat—Hermit Crabs—Cocoa-nut Planting—Dragon-Flies—Water Beetles—Jungle Spiders—Carpenter-Bee and Mason-Wasp—Eulima and Stilifer—Alligators—Mollusca—Feather-Stars—Nudibranchs—Mantle-cutting Doris—Land Shells—Reef at Pulo Pappan—Dendractinia—Weather at Labuan—Luminous Fungi. 181

CHAPTER XIII.

SARAWAK.

Entrance to River—Antimony Anchorage—Tarnuh-puti—Drift-wood—Town of Kuching—Former Condition of Sarawak—Sir James Brooke—Prospects of the Settlement—The Tuan Muda—The Dyaks—Their Superstitions—Miss Burdett Coutts' Estate—Gambier planting—Flying Squirrel—Flying Lizard—Flying Foxes—Vegetable Productions—Rain—Italian Naturalists—Quadrupeds—Domestic Animals—Dyed Fowls—Reef at Pulo Marundum . . 201

CHAPTER XIV.

THE SARAWAK RIVER.

Eclipse of the Moon—Boats and Rowers—First Halt—Reach the Rapids—The Datu and Chief Hadji—Diamond-washing—Gold—The Last Rapid—Dress of the Dyaks—The Council—Scenery of the River—Mode of Producing Fire—Journey Continued—Incidents—Change Prahu for Canoes—Return Down Rapids in the Dark—Bivouac—Malay Boat-songs—Limestone Cavern—Berlidah—Ascent of Peninjau—Dyak Village of Serambo—Rajah's Summer Residence—Bombok—Return to Sarawak 220

CHAPTER XV.

SINGAPORE.

Variety of Life in Singapore—The Malays—Their Villages—The Klings—Kling Women — Their Occupations — Religious Ceremonies — Mosque—The Chinese—The Bugis—Residences—Native Streets—Tigers, not numerous—Fire-Flies — Botanic Gardens—Sensitive Plants—Kling Bird-Catchers—Climate of Singapore—Productions of the Sea Shore—Sharks 242

CHAPTER XVI.

CULTIVATION IN SINGAPORE.

Climate of Singapore—Soil—Nutmeg Planting—Appearance of the Tree—Over-Manuring—The Nutmeg Disease—Its Causes—Ruin of the Planters — Occasional Spontaneous Recovery — Cotton—Coffee—Cinnamon—Sugar-Cane—Gutta-percha—Gamboge—Gambier and Pepper—Fruit Trees—Cocoa-nut—The Cocoa-nut Beetles—Sago Plantations 260

CHAPTER XVII.

JOHORE AND THE STRAITS.

Excursion to Tanjong Putri—Chinese Carnival—The Tumonggong—Sing-songs—Chinese Thespians—Gambling Parties—The Game of "Poh"—Gambling in Singapore and Hong Kong—Mountebank Dentistry—Opium Smoking—Statistics of Consumption—Value of Imports—Chinese Opium—Considerations—Saw Mills—Horsburgh Lighthouse—Coast of Johore—Habits of the Pill-Crab—Ubiquity of Ants 275

CHAPTER XVIII.

MANILLA.

Appearance of the City—Manilla Bay—The Town—Chinese Shops—Aspect of the Mestizas—Dilapidated Condition of City—The Great Earthquake of 1863—Features of the Shocks—Their Effects—Moral Effect on the People—Game-Cocks—The River Pasig—Tobacco Manufacture—Taxes on Commerce—Sea Snakes—Tropical Skies compared with Northern—The Southern Cross—Effects of Clear Atmosphere—Moon-blindness—Case 293

CHAPTER XIX.

HONG KONG—CHINESE NEW YEAR, ETC.

Chinese Pyrotechny—Salutations by Crackers—Religious Ceremonies—Holiday-making—Family Groups—Children—Visits of Ceremony—Boats—Toy-makers—Mandarin Processions in Canton—Irruption of Beggars—Chinese Tame Birds—Shantung Lark—Tumblers—Canaries—Mina—Street Robbery in Hong Kong—Insecurity of the Person—Police Regulations—Contrast with Canton—Character of the Chinese—Facility of Escape to Canton 311

CHAPTER XX.

CANTON.

Strangeness of Canton—Bogue Forts—Whampoa—Pagodas—Approach to City—Boat Population - Pic-nic Boats—Streets of Canton—Chops—Puntinqua's Garden—Fa-tee Nurseries—Gold-Fish—Deformities—Diet of Chinese—Dog-eating—Salt Monopoly—Unity of Chinese People—Its Causes—Insurrectionary Movements—Influence of Western Civilization—Benefits of Western Trade—Pekin Memorial on Western Education—Proposed Introduction of Railways—Language the Great Barrier—Prospects of Christianity . . 330

CHAPTER XXI.

THE SURFACE POPULATION OF THE OCEAN.

Floating Animals—Capriciousness of their Appearance—Calms—The Towing Net—Medusæ—Noctural Animals—Formosa Channel—Hydrozoa—Yellow Fly—Blue Animals in Deep Sea—Abundance of Animals in Bad Weather—Lucernarian Jelly-fishes—Their Vast Numbers—Peculiarities—Portuguese Man-of-War—Stinging Powers—Fish Sheltering in their Threads—Sargasso Sea—Its Inhabitants—Atlantic Calms—Compound Salpæ—Three Forms—Chains of Salpæ 352

CHAPTER XXII.

OBSERVATIONS AT SEA.

Flying-fish—Their Range—Object of their Flight—Always away from the Ship—Mode of Flight—Absence of Vibration of Wings—Nature of Impulse—A Flying-fish Hunt—Albicores—Abundance of Flying-fish—Trichodesmium, or Sea Dust—Red Sea Couferva—Abundance of Conferva in the China Sea—Its Range—Cases of Red Discoloration—Microscopic Characters of Sea Dust—Oscillatoria—Observations of Former Voyagers—Horizontal Rainbow—Development and Peculiarities—Changing Aspect of the Sea—Natural Colour of the Deep Sea—Changes in Shallow Water—By Rough Weather—Father Secchi's Spectroscopic Observations 373

CHAPTER XXIII.

THE LUMINOSITY OF THE SEA.

Nature of the Phenomenon—*Phosphorescence* a Misnomer—Classification of Luminous Phenomena—Sparks always visible—Their Cause—Luminous Sheath to Ship—Singapore Harbour—Simon's Bay—Noctilucæ—Scene on the Chinese Coast—Moon-shaped Patches of Light—Not caused by Medusæ—Often spontaneous—Probably Pyrosomas—Recurrent Flashes—Colour and Appearance spontaneous—Depth of the Animals—Examples of Recurrence—Milky Sea—Its Rarity—Conditions of Luminosity—Non-luminous Animals—Rationale of Luminosity—A Correlative of some other Force—Contractility—Luminous Envelopes—Range of Luminosity among Animals 391

CHAPTER XXIV.

THE VOYAGE HOME.

Storm at Hong Kong—Loss of the "Osprey"—Sea-birds at the Cape—Simon's Bay—Cormorants—Botany of the Cape—Physical Features of False Bay—Cape Town—Marine Animals of Simon's Bay—Coast of St. Helena—James' Town—Napoleon's Tomb—Ascension—General Features—Craters—Vegetation—Insects—"Wide-awake Fair"—Boldness of the Birds—Turtle Ponds—Varieties of Turtle—Western Isles—Pico—Fayal—Villa de Horta—Character of Vegetation—Spithead—Conclusion 410

APPENDIX.

Sau-o Vocabulary—Dialect of Ke-lung 434

LIST OF ILLUSTRATIONS.

	PAGE
SANDSTONE PILLARS, SOUTH SIDE KE-LUNG HARBOUR, FORMOSA	*Frontispiece*
ENTRANCE TO THE HARBOUR OF TA-KAU	35
THE SULPHUR SPRINGS NEAR TAM-SUY	70
COMATULA	137
THE WATERSPOUT	145
GROUP OF NUDIBRANCHIATE MOLLUSCA	*To Face* 195
MALAY HOUSES AT THE ANCHORAGE, SARAWAK	201
THE PANGAH, OR HEAD-HOUSE, BOMBOK	220
CHAINS OF SALPÆ	*To Face* 372
BEROÖID CILIOGRADE, FROM THE ATLANTIC	410

RAMBLES OF A NATURALIST.

CHAPTER I.

THE VOYAGE TO CHINA OVERLAND—IMPRESSIONS OF HONG KONG.

Dissolving Views—Marseilles—The Bear Rock—Flocks of Cranes—Alexandria—The Delta—Grand Cairo—The Desert—Red Sea—Aden—First Shore Hunt—Point de Galle—Tropical Calm—Lightning—Tropic Birds—Singapore—Traveller's Tree—Caricature Plant—Approach to Hong Kong—Appearance from the Sea—Boats and their occupants.

WHEN a man leaves his home and country for the purpose of making a closer acquaintance with Nature and natural phenomena in distant lands, he naturally does not wait to begin his observations until he shall have travelled a certain number of thousands of miles. And thus, although my destination was China, I found much that was striking and interesting, from the point of view which I had chosen, on the road. The overland route to India and China is indeed so generally known, and so experimentally familiar to a large number of persons, that it would serve no purpose to dwell upon its details; but nevertheless a work, whose plan mainly seeks to recount the aspects of Nature in foreign countries, would scarcely be complete were no reference made to so large a portion of travel as is passed over by the steamers of the Peninsular and Oriental Company between Marseilles and Hong Kong.

The great drawback of this route, to a person not travelling on business which requires despatch, is the restless rapidity of movement which allows of no quiet, except on the calm days at sea. When at length, after an interval of a few such days, land is reached, he catches a glimpse of a country, it may be the most interesting he has ever visited; but in a few hours, almost before he can realize that it is not a pleasant dream, inexorable necessity attracts him once more to the ship, and he turns his back upon the new country, it may be for ever—its people, its vegetation, its scenery, leaving the impression of an unreal vision upon his memory, which will endure as such as long as he lives.

I can only compare the passage overland to the picture patterns seen in a kaleidoscope, changing with such rapidity that the impression of one is still vivid when it is succeeded by another, while yet each picture is complete in itself, and has features which distinguish it no less from its predecessors than from its successors. Or I might liken it to a series of dissolving views, in which the eye still dwelling upon the last and recently formed picture, finds it replaced by some other and contrasting one, when another port is reached, which gradually, by the force of its present reality, drives out of the mind (for the time) the one which has for some days occupied all its thoughts. Thus we change the verdurous Delta for the arid desert plain—and this again for the piled-up, barren rocks of Arabia—which in turn give place to the green and smiling fertility of the palm groves of Ceylon, and the scarcely less luxuriant islands of Penang and Singapore.

The vine-clad hills and olive groves of the south of France, and the tall flower stems of the great thick-leaved aloes about Marseilles, were hailed as some foretaste of that luxuriance of vegetation which we were soon to witness in the tropics, of which, however, we were destined to see but little until we had reached its perfection in the rich garden of Ceylon. The rocky islands of Corsica and Sardinia, the picturesque Maritimo, and the white fortifications of Malta, present nowhere a tree or a bush to relieve their barrenness, and the only thing remotely interesting to a zöologist is the semblance of a gigantic bear which may be seen walking down a shelving crag in the straits of Bonifacio, and appropriately called the Bear Rock. At Malta, some hours of moonlight and dawn were all that were allowed us, which thus prevented landing.

Steaming down the placid waters of the Mediterranean at this time of year (the beginning of March) is perhaps the perfection of pleasant travelling. The air is more soft and balmy than it can be at any season in our own climate, and without being oppressively warm, all thought of cold is abandoned, and a sensation of agreeable exhilaration mingles with one of ideal comfort. Once or twice a flying fish broke the quiet surface of the water, but the Mediterranean species did not make their appearance in any numbers. South of Candia my attention was arrested by four successive flocks of cranes flying northward, whose screams could be heard distinctly as they passed close over the ship—a circumstance which was stated to be unusual, the captain of the ship never having observed it before in his frequent passages up and down the Mediterranean. There were about 50 birds in each flock. The first assumed a long irregular line, but the other three were more or less wedge-shaped, particularly one, in which the figure was remarkably symmetrical. Buffon's idea of the cause of this peculiar arrangement, viz. that the strongest naturally keeps first, while the rest are necessitated to follow behind, would hardly, I think, account for the regularity of the figure, which is maintained until it faints into the tenuity of a spider's thread in the distance. His theory is, indeed, no improvement upon that of Cicero,* who, being an augur, studied the flight of birds, and supposed that when migrating in large bodies, they assumed the ▷ form from an intuitive knowledge that it offers less resistance to their rapid flight—while one, probably a strong-pinioned bird, is selected by them as their leader; though it is probable that this one, in turn, gives place to others during their progress.

The first sight of the shores of the Nile Delta is by no means striking, though just what a consideration of its nature would lead one to expect. A long low coast, terminating in sand-hills, is presented to view, whose monotony is only broken by distant and somewhat formal rows of date-palms, interspersed with windmills. But here, as we land at Alexandria, our first dissolving view fairly gives place to a totally new picture, in which the colours of the kaleidoscope play a conspicuous part. We seem at once plunged into the embodiment of the dreams of those days when we read of Aladdin, The Three Calenders, and Haroun-al-Raschid. Streets narrow and winding—shops open to the street, and without windows of any kind—merchandise piled up around the owner, who sat cross-legged upon the counter, smoking his pipe and awaiting custom—barbers shaving their customers in public—divans, where Arabs were sipping coffee meditatively—bakers and provision-merchants with wares anything but tempting to a European—and Nubians, as black as jet, carrying water-skins, of which the outsides were sufficiently disgusting objects. Passing to and fro through

* Cicero, de Naturâ Deorum, lib. ii. cap. 49.

the streets, which were impregnated with a penetrating and far from agreeable odour, and under a sun more powerful than we had ever felt in England, were crowds of Arabs, Turks, Negroes, and Egyptians, in every variety of costume and of every shade of colour;—some in gaudy dresses, and flowing robes, turbans, and fezzes—others with bare legs and arms, or wearing only a kind of smock, and variously contrived head-coverings—women veiled, with a long strip of black hanging down from their eyes to their feet; others in dazzling white, veil included, or with little ones astride upon their shoulders—children of all ages and degrees of dirtiness, crying for *backsheesh*—lean mules and donkeys, with turbaned and portly Turks upon their backs, or with bundles of merchandize hanging on either side—strings of camels laden with various goods—warlike gentlemen with long curved scymitars at their sides, and pistols two or three feet in length stuck in their belts—all these together formed a combination often described, perhaps, but not to be forgotten when once seen.

A land journey is always a pleasant interlude in a long sea voyage, though the small carriages of the Pasha's railway, generally filled with their full complement, are not the most delightful of conveyances under an Egyptian sun. Across the fertile Delta agricultural operations were everywhere going on. Groups of date-palms (Phœnix dactylifera) and groves of olive trees constantly met the eye; numerous camels, herds of buffaloes, mingled with the coloured domestic cattle, goats of a small size, and broad-tailed sheep, were in plenty throughout the route. The fields were often separated by hedges of mimosa (M. Nilotica), and frequent villages occurred, mere collections of mud huts, squalid and desolate (and sometimes deserted and in ruins), about which were Arabs equally squalid and wretched-looking, the children cased in dirt, wearing a single scanty and dirty garment, with eyes more or less affected with ophthalmia, and holding out their hands for *backsheesh* as we passed. Villainous-looking dogs, like gaunt jackals, lurked about the huts, and luxuriant cacti flourished in some of the gardens. Large kingfishers hovered over the ponds, and handsome black and white stork-like birds stood motionless like sentinels by the side of canals—buzzard-like hawks flew familiarly about, and occasionally swooped down almost among the people collected at a station—crows, and plovers, and sparrows, were not uncommon, particularly the last, probably, however, not our domestic species, but the tree sparrow (Passer montanus).

As the sun went down, the zodiacal light appeared very distinctly; and for several nights I remarked it as we passed down the Red Sea, much more clearly than I had ever observed it in England. Conspicuously upon our right hand shone out the Egyptian star, Canopus, never visible in this latitude; but whose first sight roused associations in unison with the classic locality we were traversing,—for at sunset we had crossed a branch of the Nile.

The glimpse afforded by a ride through Cairo did not differ essentially from that described at Alexandria; but the city is far more interesting and remarkable; the streets more narrow and mazy, ornamented with arabesques and frescos; minarets and domes meet one at every turn, while the people seemed even more essentially Eastern than in the commercial town of Alexandria. Giving my ass his reins I diverged from my party, and let the beast take me where he would, trusting to his instinct to lead me finally aright; and thus unencumbered, I could gaze at my ease upon the motley crowd of well-conditioned Turks and white-veiled ladies, running Arabs and sooty Nubians,—mules, donkeys, and camels, which threatened to overturn the crowds of active, half-naked, and dusky children which ran hither and thither across the streets. Having thus paid a visit to the Mosque of Mehemet Ali, and seen its marble pavement and pillars of alabaster, and having gained a distant view of the Pyramids, another scene succeeded the vanishing picture of Grand Cairo.

This time it was the desert; barren sands and low stony hills of an uniform and monotonous brownish-yellow tint, broken here and there by a stunted vegetation—these small green oases being, however, few and far between. But little life is visible here. A few black crows flew about the outskirts of Cairo and Suez, but did not penetrate far into the desert, although some mud villages of the Arabs made their appearance even here; and we more than once passed a group of Arabs, accompanied by their jackal-looking dogs, apparently walking through the desert along the line of railroad. Now and then the skeleton of a camel lay bleaching on the ground, more particularly in the neighbourhood of Suez, and at the central station of Awebed a small lady-bird (Coccinella) flew into the carriage. It was a true desert species, of the characteristic pale brown isabelline colour, admirably matching the prevailing tint of the sands on which its lot was cast.

The passage down the Red Sea was cool and pleasant, but uneventful. An agreeable breeze followed us, and favourable contrasts were drawn by many with previous experiences. Passing Mocha, we were near enough to see its white houses, over which the unusual phenomenon of rain was falling abundantly, and it was also raining further to the south, over the land. Barren rocks accompanied us the whole way, beginning with the distant mountains of Sinai and Horeb; and having left behind the volcanic islands of Zebayer and the peaks of Babel-Mandeb, to both of which we were quite sufficiently near to discern their evidently crateriform character, another picture was for a few hours presented to our bodily eyes. This time it was the barren, craggy, extinct volcano which is now Aden, where the scene was as wholly new and distinct as any previous one. The black Somali from the opposite African coast, their heads either plastered with chenam, or their light-yellowish hair, dyed from its native hue by this treatment, and woven into long ringlets all over the face, here mingled with Arabs, and formed fitting denizens of a country that was little better than an arid desert. Long strings of camels toiled in procession up the hills, laden with water-skins, fire-wood, and bars of iron,—and here and there a half-naked negro met us, seated upon a dromedary's hump, and passing us at a long swinging trot, such as only a camel could accomplish. Vegetation was here scarcely less rare than in the desert itself—small patches of green, however, were here and there visible, produced by a Resedaceous plant (allied to mignonette), which struggled to maintain existence—and it was not until we arrived at the neighbourhood of the great water tanks that we observed how the industry of man had converted a wilderness into a garden, and at infinite labour and expense had not only conveyed thither flowers and plants from distant regions, but even the very soil in which they were growing. Here, in the yawning mouth of what, ages back, had been the fiery gulf of a great volcano, but of which nothing but the form now remains, are the cantonments or military stations, and all around is life and bustle.

But time, and the unwonted and intense heat, would not permit of more than a cursory view of those great works, the water tanks, or of the curious scenes of Eastern life which are to be viewed in the great market-place and bazaar; and having taken our glimpse we returned as we had come.

Having a short time at my disposal before rejoining the vessel, I went down to the beach, where, although the water was rather high, I met with some matters of interest. Under the stones on which were many largish Chitons, were numerous grape-like eggs of the cuttle-fish (Sepia), each egg containing a small well-developed cuttle, which, when detached, at once moved actively away, and discharged ink from its ink-bag. I was fortunate in finding, also, under a stone, three specimens of a beautiful nudibranch, or marine slug, which I kept alive for some days for the admiration of some intelligent fellow-passengers, who expressed their astonishment that such brilliantly-coloured and graceful creatures should exist, and many were the questions as to how I had found them. They were of the genus Bornella, and probably *Bornella digitata* of Adams, a rare species, which had only previously been met with in the Straits of Sunda by Mr. Adams (in H.M.S. "Samarang"), two or three specimens, and the same number on the Madras coast. Delicately marbled with vermilion streaks, they swam freely in the water by a lateral twisting movement of the body, waving at the same time their singularly complex and elegant tufts in a most striking and graceful manner. Vain, however, were all my attempts to depict these nudibranchs in a satisfactory manner, for the conditions on board a mail-steamer are by no means favourable for such studies. I therefore placed them in glycerine, which has a wonderful power of retaining the bright colours, transparency, and delicate outline of some of these perishable animals; but has, unfortunately, the drawback that it unfits them for subsequent dissection, so that it is always advisable to place some specimens in this medium, and others in spirit, for the use of the comparative anatomist.

Once more, after a week's voyaging over the calm waters of the Indian Ocean, the view changes, and for a few hours we are walking through the cocoa-nut groves and cinnamon gardens of Ceylon. Glad, indeed, were we of the shade afforded by the over-arching palms, which here, for the first time, greeted our eyes with all the luxuriance of equatorial vegetation,—a change rendered the more agreeable and striking by the contrast it afforded to the barren rocks, which, since we quitted Marseilles, had everywhere met our view, excepting only the green patch of Delta between Alexandria and Cairo. Here plantains and pumilows, limes and pine-apples are to be had almost for the asking; and here, after a glorious drive through a forest of palms, thickly studded with native cottages, about which dusky forms hovered, and little naked children who required no protection for their tender bodies, we at length seated ourselves beside a bed of the sensitive mimosa, and enjoyed a prospect as though the view from Richmond Hill had been transported to a tropical clime, with all the voluptuous accompaniments of a garden in Paradise.

But in these latitudes during the fine season the ocean presents aspects nothing inferior in glory and magnificence to the scenes beheld on land. A perfect calm, such as occurred a few days later, was a thing to be remembered; and although I have seen many calms since, they have by no means always combined every element of beauty which tended to make this one unique. The sea was like an azure mirror, polished, spotless, and brilliant, in which the slightest mote would have seemed a flaw; but from out of which, from time to time, shoals of flying fishes, like flocks of little white birds, emerged, with a splash and a whirr like a covey of partridges, dropping one by one into the water again like a shower of canister or grape, and leaving only a few ripples which presently subsided, and the water was once more like a clear sapphire. The sky was filled with noble cumuli of various shades of white, arranged in successive piles or layers from the zenith to the horizon, flat below, massively rolling above; and so crystal-clear was the atmosphere that those most distant were as well defined as those nearly overhead; and even the clouds below the horizon, and of which only the flocculent convoluted tops were visible, were sharply cut against the distant sea-line. It was like a noble temple, whose floor was lapis-lazuli, and whose roof was infinity. But once before had I witnessed a parallel scene, but with the colours reversed, when far up the recesses of Mont Blanc, the deep unwonted blue of the cloudless sky was cut by the clear, trenchant outlines of spotless aiguilles which towered up all round from the pure white floor of the snowy glacier.

Events interesting to the observant naturalist can hardly fail to happen each day while traversing the ocean, and it is not to be supposed that during this time nothing was seen worth recording; but I have thought it better to collect the various circumstances worthy of notice in a separate chapter, on the surface life of the ocean, than to speak of them in a piecemeal and isolated manner, which also would stand in the way of any interesting generalisations. Scarcely a day passes, however, without some addition to one's stock of observation and information, whether it be a fish swimming the sea, a bird winging the air, or some floating delicate animal which would seem least fitted to buffet with the waves, which at some seasons lash themselves into irresistible fury.

The straits of Malacca, with its fine prospects of Sumatra, gave a taste of those tropical storms which have procured for them the name of *Straits' weather*. Not that we were at any time involved in the thunder cloud; but on this and on several other occasions, certain peculiarities of electrical phenomena occurred, which may be appropriately referred to here. In the first place, it has always struck me as a singular phenomenon, that day after day thunderstorms have apparently been bursting around us, in several places illuminating the horizon, and yet we seemed to be exempted from them. This was particularly the case in the Straits of Malacca, and on the coast of China. Nor was it all of the kind known as summer lightning, for I have delighted to watch the vivid spark coursing through the air, or dashing down upon the sea or land; but although I have so often watched lightning night after night successively, the sound of thunder has been a rare occurrence.

Again, on two occasions I have witnessed storms which have apparently been of such severity that to be situated beneath them must have seemed like being at the mouth of hell. Once at Shanghai, in July, the sky was illuminated with one incessant unintermitting glare, lasting several hours, but no thunder was heard; and a similar circumstance took place in May off the south coast of Madagascar, when a storm broke to the south of us, even exceeding this in grandeur. From 7 to 11 P.M. a flickering glare, which left the sky dark only for a second once in half an hour or an hour, showed that a terrific elemental strife was going on. The central point seemed elevated 10° or 15° above the horizon, and as the nearer clouds cleared away I watched for hours the unceasing flashes—tongues of fire darting out round the distant clouds—radiating in five or six distinct streams of flame from a given point, like the thunderbolt in the hands of Jupiter—coursing along the sky, or dashing down into the sea at the horizon like liquid fire; but all this while not a sound was heard, no thunder reached the ears, and the position of the storm scarcely altered during the whole time. At 11 I retired from watching it, but as long as I remained awake I could see the reflection playing upon the walls of the cabin, like the flickering of an unsteady candle. The day succeeding was marked by a brilliancy of atmosphere and freshness of temperature we had not experienced before, and the only important change we observed was an adverse wind.

On this occasion I noticed a peculiarity, which was also very strikingly marked, in a storm which passed near us at Sarawak. The lightning in this case was unusually vivid, but the flashes did not have the appearance of simple instantaneous sparks, but looked just as though they consisted of liquid fire poured out from a vessel in a continuous stream, and lasting a perceptible time, during which the lightning vibrated upon the retina—the zigzag form of the flash, however, being perfectly retained meanwhile.

As we neared the Straits we observed several floating logs, or trunks of trees, which in the distance looked like boats. Some of them were covered with gannets (Sula alba?), as thick as they could cluster, though the birds could have rested, if they chose, one would suppose, upon the water, belonging as they do to the fully-webbed Pelicanidæ. These birds are seldom seen far from land, and their appearance is a sign of its proximity; not so, however, with the Tropic birds (Phaëthon æthereus), beautiful black and white creatures with yellow beaks, and conspicuous for their long pointed tails. Four of these birds appeared about the ship on two successive days in the Indian Ocean, on the second of which we were 800 miles from land. As long as I was able to watch them I did not see them settle upon the water, nor did they appear to attempt to catch the flying fishes, which at the time were on the wing in considerable numbers.

An hour's walk in Penang gave the first glimpse of Chinese life, and one could not fail to be struck with the activity and energy displayed here as everywhere by the celestial race,—all astir and busy, though but just daylight. The verdure of the place, and the elegance and grace of the various specimens of palms which met the view, were an agreeable relief to the eye after a week at sea. The dense jungles and sandy beaches of the Malacca peninsula were visible as we proceeded towards Singapore, and at night many fires were visible, which made one speculate on the occupation and characters of the inhabitants of this tiger-haunted land. At Singapore, after threading the green, wooded islets which conduct to the harbour, having ridden through hedges of bamboo, groves of cocoa and betel-nut, mangrove swamps on which were built villages, forcibly recalling to mind the ancient lake-habitations,—streets peopled by Malays, Chinese, and Klings or Madrasees in every variety of picturesque costume, I at length found myself in the verandah of a bungalow, and overlooking a garden in which many strange trees and plants were growing. Among these was the Traveller's tree (Urania speciosa), the banana-like leaves of which spring in a beautifully imbricated fashion from the two opposite sides only of the stem, the whole tree representing a gigantic open fan. The rain falling upon the leaves and leaf-stalks, runs downs a channel in the latter until it reaches the base, where a reservoir is formed by the sheathing petioles, which so closely embrace one another that it cannot escape. An incision, therefore, through these sheaths produces a constant fountain of pure, refreshing fluid, of which the experienced traveller may at his pleasure avail himself. Another singular tree, or rather shrub, I first observed here was commonly known as the face-leafed plant, or Caricature plant of the East Indies (Justicia picta), every leaf of which exhibited upon its blotched surface a series of remarkable caricature resemblances of the human face divine. One of these trees in the garden of Gustave Doré would be worth a fortune to him, supplying him with a never-failing fund of grotesque physiognomy, from which he might illustrate every serio-comic romance ever written by Swift or Dickens, by Rabelais or Cervantes. About the verandah the most common bird appeared to be the rice bird, or Java sparrow (Loxia oryzovora); but on a subsequent occasion, in November, I looked in vain for these birds, which had been so plentiful in the beginning of April.

I shall, however, have occasion to return to Singapore, and shall therefore now proceed on our journey across the China Sea, at this season beautifully calm; and another week of delightful *dolce far niente* brought us near the goal to which for six weeks we had been constantly travelling. The day before reaching our ultimate destination of Hong Kong, we experienced, for the first time during all the voyage, squally and unpleasant weather, which was not disagreeable, however, when regarded simply as a change from the uniformly fine and calm seas we had experienced for six weeks; more particularly as there were no signs of the dreaded typhoon. As we approached the coast, great numbers of junks, with mat sails and two masts, appeared, the high poops of which gave them the strange aspect of plunging headlong into the water; but they appear to be excellent sailors, and under ordinary circumstances have no real tendency to do so. The numerous islands clustered about the entrance of the Canton river began to make their appearance on the following morning, bare of trees, but usually smooth and more or less green; and ultimately the back of the island of Hong Kong itself, sparsely dotted with handsome residences, though otherwise not very prepossessing, being barren and exposed, interlaced by craggy ravines, and running up into elevated crags, the highest of which, surmounted by a flagstaff, is called Victoria Peak. This is the telegraph station, from which the appearance of every ship that approaches the harbour is signalled, and from which a booming gun announces to every expectant inhabitant of Hong Kong, and to every ship in the port, that the mail is in sight.

The first sight of the Hong Kong of the present day is something not to be forgotten, and perhaps unequalled by any view of the same character. Having passed Green Island, a round knoll in mid-channel, we begin to sight the shipping, from the midst of which a puff of smoke announces that the "Fort William," Peninsular and Oriental receiving ship, now sights us rounding the angle of rock. On our left is the long stretch of sea ending in the Capsingmoon pass, through which lies the way to Canton. The rugged crag of the Peak rises on our right, at the base of which the town lies like a city of palaces, gradually developing as we proceed round the front of it. Meanwhile our attention becomes divided between the varied and numerous shipping, and the magnificent scene which the land affords on either hand, but more especially upon the island side. Passing by a heterogeneous crowd of junks which ply between Hong Kong and Canton or Macao, for the supply of the market, and which attract notice from their novel and foreign aspect, we thread our way among British shipping of every class, among which also are many British men-of-war and gunboats, mingled with not a few foreign vessels. Through these, often within speaking distance, we steam slowly and cautiously, and have time to watch the gradual unfolding of the city of Hong Kong. Built terrace upon terrace up the base of the hill, a series of splendid palatial residences, with open verandahs around them, rise nobly, and in strong contrast to the dark back-ground of the craggy peak which towers above them—on one side sloping gradually down towards the Lyemoon pass in the East, and on the other side suddenly and precipitously terminating towards the sea in the West,—reminding the spectator very strongly of the Rock of Gibraltar. On the opposite shore, a long line of barren, serrated peaks, sweeping picturesquely up from the shore, with nowhere any sign of life or habitation, forms a striking contrast to the busy life and activity of that from which we have just turned.

Although from the nature of the ground there are numerous spots on shore which command exquisite views, the position of the spectator on ship-board in the harbour is peculiarly adapted for obtaining the most charming and picturesque scenery—for as the vessel swings with the tide, a series of panoramic pictures, as it were, is gained, embracing every quarter of the compass; and these, when seen from so advantageous a position as the stern galleries of the

"Princess Charlotte" in fine spring weather, were truly delightful. And when night came, and the young moon lighted up the scene, the rows of lights round the dark mass of the mountain, which itself stands out in bold relief against the twilight sky—the smooth, bright sea reflecting the moon-lit heaven, and bearing upon its surface innumerable sombre ships, each showing its guardian lamp—was a fairy-like scene which I never tired of watching.

Without leaving the ship, too, there were points of interest in Chinese life which forced themselves upon the attention. The numerous boats, or *sampans*, plying about between the shore and the various ships, all manned by Chinese, were in themselves a study; and some were constantly hanging about the ship at a respectful distance, in the hope of a fare. They are, for the most part, long boats with a small awning near the stern, under which the passengers sit, and they have a complement of four or five rowers. I have said they were *manned;* but although there are usually one or two men among them, the majority of the rowers are women, or young girls. In fact each boat is the home of a family, and in their boat they spend their whole existence—how, it is difficult to comprehend; but naturally they become expert in rowing and handling their craft. In fine weather but little skill is required, but there are times when the sea in the harbour is so rough that boats cannot be obtained at any price; and I have heard of people, only last winter, being detained on board ship for nine days together, unable to get ashore. The family inhabiting a boat all share in the work—it may be a husband, wife, daughter, and son—or, if the family is not sufficiently numerous, the complement is made up by agreement from without. But the women and girls, whose dress differs but little from that of the men (when the latter wear any), are no less strong and active than their lords; and, moreover, they are not unfrequently burdened in a manner which would at first sight seem to hamper their movements very considerably. Strapped upon their backs, it is a common sight to see an infant, his little bare feet peeping out on either side, and the unsupported head tumbling from side to side with every movement of the mother, who, in the act of rowing, places herself in postures by no means always suitable to the child's comfort; but it seldom complains, and seems to become accustomed to the strange rocking motion. Often the mother hands the infant over to the back of a child, girl or boy alike, of nine or ten years old, who moves about the boat apparently with little reference to his burthen. Children of an age to toddle about, but still so young as to require attention, are often tethered by a string to the middle of the boat, or ornamented by sundry gourds fastened to their bodies, so that in case they fall overboard, as I have seen them do, they may float until picked up.

The numerous cargo boats plying in the harbour add to the liveliness of the scene. They are mostly rowed by men, who stand up and push the oar before them. A strange sight it is in wet weather to see these men, who, under these circumstances, wear cloaks made of grass,—the raw material sewn together,—which, crowned by the broad and pointed bamboo hat, give them an aspect of savagery which can scarcely be surpassed.

CHAPTER II.

HONG KONG TO PRATAS ISLAND.

Anglified appearance of Hong Kong—Physical Character—The Chinese Quarter—Leave Hong Kong—Arrive at Pratas Reef—Description of the Island—We visit the Island—Its Vegetation—Insect Fauna—Marine Animals—Sea Weeds—Fishermen's Temple—Lagoon—Birds of the Island—The Gannets' Settlement—The Seine—Towing Net—Rollers—We quit the Reef—Birds observed upon the Ship between Pratas and Formosa.

It scarcely forms a part of my plan to enter largely into a description of Hong Kong, nor shall I attempt to do more in this place than cursorily refer to some of those features which most strike a stranger from the West. I shall have occasion to return again to the island, which is a convenient starting-point for many places.

Hong Kong is so essentially English China, that a traveller who passed by here, and visited no other part of the country, would have but a very imperfect idea of Chinese life and manners. The houses are fine, substantial, and European for the most part, the Chinese town forming quite a subordinate part of the place; and the population is a mixture of English, French, Portuguese, Americans, Parsees, Mahommedans, and Chinese. Of these, all, except the Chinese, are of a good class, being for the most part well-to-do merchants, who employ the Chinese in their offices either as compradores, clerks, servants, or coolies. Some Chinese there are who do business in Hong Kong on their own account, and the compradores of the large European houses are often highly respectable men; but the majority of the Chinese population are of an inferior class. Nearly all of them understand enough English to carry on a tolerably free intercourse with their masters, though this English is of that mongrel kind known as *pidgin* (or business, pronounced by them *bidgness*) English, which it is not only necessary to understand as spoken by them, but also to speak freely in order to be intelligible to the Chinese.

The island is syenitic granite, of a kind which very readily decomposes upon the surface where exposed to the weather, and the water which percolates from this disintegrated rock appears to have deleterious properties, to which the unhealthiness of some parts of the island would seem to be mainly due. Irruptions of trap are visible in some parts, and the whole island partakes of the characteristic barren aspect of the greater part of the Chinese coast, and, except in sheltered situations, as in the part called the Happy Valley, is for the most part destitute of any trees, except a stunted pine. Good roads are constructed round the greater part of the island, often high up the hill side, which command glorious prospects over the sea, and the rocky and elevated mainland of China; which, with the ever varying appearance of the harbour crowded with shipping of every nation, render a walk upon the upper road one of the most picturesque and grand that can be anywhere met with.

There are no Chinese features, however, observable about Hong Kong, which are not seen better in other and less hybrid parts of China. Everything is more or less diluted with the European element; and I was much struck, when I first observed a small-footed Chinese woman of superior class meandering with painful steps through the street, accompanied by an elderly attendant, to see that the sight seemed to attract as much attention from the Chinese population as it did from myself. Around every shop door were clustered curious groups, who watched the fair hoofed lady until she was almost out of sight, though I imagine their curiosity was chiefly excited by the unusual appearance of a lady so evidently superior, walking in the streets. Still there was much that was curious, and could not fail to interest me; and in order to observe them more at home, and free from foreign interruption, I bent my steps into the purely Chinese quarter, where I soon made the discovery, however, that squalor and dirt and crime were also here at home. How I suffered for my temerity I will recount in another place; suffice it to say that here I learned a lesson I did not forget as long as I remained among the Chinese people.

After a fortnight spent in Hong Kong, I joined her Majesty's despatch boat "Serpent," Commander Bullock, who most kindly shared his own accommodation with me. The destination of the "Serpent" was the Formosa Channel, with a probability of visiting some of the ports of the island of Formosa, the coast of China, and Shanghai—a hope that was fully realised by the event; while the various delays, and the devious course rendered necessary by surveying operations, gave me much and desirable opportunity of prosecuting my observations. We quitted Hong Kong on the 24th April, steering first south for the Pratas Island, which we reached on the 28th.

Pratas Island is situated in lat. 20° 42′ N., and long. 116° 43′ E., and is of a horse-shoe shape, occupying the centre of the sunken or western part of the great Pratas reef. The reef itself is of a crescentic form, extending 13 miles to the eastward, and having a breadth from north to south of 12 miles, enclosing a lagoon of about 10 miles in diameter, dotted over with numberless coral patches and shoals. It lies in the direct line of route between Manilla and Hong Kong, and is therefore a spot where many a good ship has been wrecked, especially upon its south-eastern side, which is too often concealed by the thick fogs which prevail during the north-east monsoon. The Pratas reef and island were surveyed by H.M.S. "Saracen," J. Richards master commanding, in 1858, and at that time it was believed that vessels of 15 feet draught could enter the lagoon by the south channel, between the south side of the island and the south-west horn of the reef; but in our recent visit Capt. Bullock found, that although only drawing $12\frac{1}{2}$ feet, he could not safely make the attempt, and consequently the ship was anchored on the edge of the reef, three miles south of the island, which thus sheltered it from the strong northeast wind blowing at the time.

Pratas Island is about a mile and a half long, and half-a-mile wide, and is only visible at a distance of eight or nine miles in clear weather; not rising in its highest part more than 25 or 30 feet above the level of the sea, though the bushes which cover some parts give it an additional elevation of 10 feet or so.

On Monday morning, April 30th, with Capt. Bullock and Mr. Sutton, chief engineer of the "Serpent," I visited the island, two hours' pull from the ship, and spent the day in exploring its character and natural history features. It is formed entirely of coarse coral-sand or débris, generally shelving gradually, but in some parts having a steep bank about three feet high. The interior is rough and hilly, from accumulations of similar white sand blown up from the shore, and so overgrown is it with shrubs as to be in some parts almost impenetrable, though the soil might be supposed to be anything but favourable to vegetable growth, nothing but sand being anywhere visible, and that of the coarsest and loosest description. The bushes in some places approach very near the sea, and between them and the water's edge various flowers not unfrequently peep out from the inhospitable soil, including a potentilla, an anemone, a plantago, and some grasses. On the west side of the island is a deep indentation into which the sea enters, forming a shallow lagoon or bay, on the banks of which the vegetation assumes quite a park-like aspect; bushes, and even small trees, with spreading branches springing forth close to the ground, producing a scene of great luxuriance and some beauty. Amongst the bushes immense orthopterous insects (Grylli) flew about, exhibiting a deep-red underwing, and looking very much like small birds. To the shrubs also were attached numerous geometric webs, which were occupied by a species of spider belonging to the genus *Acrosoma*, having a squarish abdomen, from the upper surface of which projected several spike-like processes. This was the only species of spider which came under my notice; and entangled in its web there appeared to be as often a spider of the same species as any other kind of insect, the paucity of insect life on the island apparently driving them to cannibalism. A moth, whose expanse of wing was about an inch, and having small red and black spots upon it, was pretty numerous, and appeared to be the only lepidopterous insect, with the exception of a large clear-winged species, which was captured, but unfortunately escaped again. These, with some ants and a few beetles, constituted the insect fauna, as far as could be determined during our single visit. The beetles were a species of Dermestes, and a little Corynætes, cosmopolitan in its habitat, and common nearly all over the world.

Among the coral-débris upon the beach were numerous masses of various sizes, consisting of rolled Astræas, Madrepores, &c.; and mingled with them were fragments of shells of a great many species of Conus, Cypræa, Turbo, Pinna, Hippopus, &c.; but none of them entire. Innumerable little hermits (Paguri and Cenobitæ) occupied the deserted shells of Naticæ and Neritinæ, and larger ones those of good-sized Turbines; but I saw no live shells upon the beach, except a few insignificant ones, such as Litorinæ and Purpuræ; nor, though the water was bright and clear, and I waded out as far as I could go, could I anywhere see traces of Annelids or Echinoderms. The harder parts of the sand were perforated with deep holes of various sizes, from which emerged from time to time a wary and swift-footed crab (Ocypoda), which scuttled nimbly down to the sea upon the first sign of approaching footsteps, and appeared to be aware of us at least at 50 yards distance. Nor was it easy to capture a specimen, for while on the one hand they never made the mistake of running *away from* the sea, on the other hand, if cut off, they fled so quickly, and *doubled* so nimbly, suddenly running the opposite way without the clumsy process of turning round, that they afforded great amusement and not a little exercise and exertion.

The sea in the neighbourhood of the Pratas Island has a very variegated appearance, from the alternations of bare white sandy bottom, with patches of Ulva and Zostera, both of which are very abundant. The Ulva is a very beautiful reticulated species (Ulva reticulata, Forsk.), and the Zostera leaves float about in all directions and in all stages of decay, generally bearing upon them minute dendritic polyzoa, orbitolites, spirorbis, &c., with which the towing-net from the ship was replenished. Besides the Ulva, I obtained several other species of seaweed, washed up on the beach, and conspicuous among them a species of Padina, very abundant everywhere in these seas, and a Sargassum.

As might be expected on so small an island, quadrupeds are scarce, nor did we observe any, though it is said the universal Rat was seen there when the "Dove" visited the spot, nor did I notice the bones of any quadrupeds which would have indicated their existence there. The skeletons of turtle were met with more than once, but whether they visit the island, or are cast up dead upon the beach, I am unable to say. No other traces of reptiles were observed.

Pratas Island is occasionally visited by Chinese fishermen, who repair to it in the early part of the year, and there is a good junk-anchorage in the north-east corner of the lagoon. We soon came upon traces of such a visit in a clear patch among the scrub, in the midst of which a well had been sunk, from which brackish water might be obtained. There were scattered about various implements of pottery, in the shape of water-vessels and teapots, some entire and others more or less broken, and surrounding them were strewed great numbers of shells, of a species of Strombus (S. Luhuanus), the remnants of a past feast, and which remained to form a future kitchen-midden in the sand. At the head of the shallow inlet or lagoon stood a joss-house, or Chinese temple, in a rather dilapidated condition from the effects of wind and weather, the roof nearly torn off, and the plank walls very shaky, so that the rain and weather had left their visible traces also upon the contents and furniture. In this rough building were 30 or 40 josses, or wooden idols, of various sizes, once resplendent in paint and gilding, but now faded and weather-worn. They were arranged symmetrically upon a sort of altar, and upon the tables before them were bundles of joss-sticks, packets of joss-papers, rouleaux of paper dollars, lucky stones, gongs, tom-toms, while around the building were grotesque wood carvings, procession staves, and all the paraphernalia of the Chinese devil-propitiators. We soon found, however, that they must be handled with caution—they were rotting with damp and decay, and harboured numbers of small scorpions, white ants, and ugly-looking spiders, which commanded a certain amount of respect from their malignant and venomous appearance. The blue-jackets especially, with their bare feet, were very shy of walking about in a spot where scorpions had their habitation, but fortunately no one suffered from their stings. Among other offerings to Joss, were a number of large model-ships, representing three-deckers, and made of paper stretched upon frames of wood, now much torn and dilapidated, but which showed plainly the piratical tendencies of the frequenters of the temple, and their desire that Joss should cast some barbarian ships upon the shore for them to plunder. As far as we could judge, however, from the condition of the place, it must have been three or four months since anyone had visited the island.

A slope of long, rank grass led down from the joss-house to the shores of the shallow inlet, upon which, and in the water, were strewed immense numbers of dead shells of Cerithium vertagus, some few of which were inhabited by hermit crabs. From observations made at the island upon the tide, it appeared that during the day of full moon it was high water at 8 a.m., and ebbed until 3.15 p.m., by which time it had fallen three feet. It was not surprising, therefore, that some of these deserted shells were high and dry; but this would hardly account for the fact that, considerably above high-water mark, many lay half-embedded in the dried mud and thick confervoid growth which had long lain above high-water mark, and bore the signs of having been well baked and cracked by many a noonday sun. The banks of the lagoon had evidently been under water comparatively recently, and much higher up than the tide now reached.

But although some classes of animals were poorly represented upon Pratas Island, there were plenty of birds, and of several species, both sea and land birds. A buzzard I noticed several times; but it was too wary to allow me to come within gunshot, although it offered a tantalising mark just out of range. I observed a very handsome shrike, with an ash-coloured head and black moustache. The bluejackets reported that they had seen a canary; and I afterwards saw myself a yellowish bird resembling the English siskin, which was probably the bird they had noticed. Another bird (Petrocinclus maniliensis), about the size of a blackbird, was of a glossy metallic blue above and fawn-coloured beneath. Its stomach contained the elytra of beetles. A fifth species presented all the appearance of a veritable blackbird, but I could not get near enough to examine it closely. A species of swallow, probably Hirundo gutturalis, with glossy bluish back, chestnut throat, and with a speckled fawn-colour underneath, was flying about in considerable numbers; and on the banks of the shallow inlet I saw a bright-coloured kingfisher, very similar in appearance and size to our own species. There were also some small birds which crossed our path from time to time, with the jerking flight and the chirrup of the hard-billed perchers. Large flocks of Tringas (sandpipers), of at least two species, were visible on the sandy flats of the inlet which were left uncovered in the afternoon, and also upon some parts of the seaward shore of the island, where it was inclined to be soft and marshy. There were also two species of plover, the one of a reddish-brown colour, with orange-red legs; the other of a delicate mouse colour, with yellow legs; and a godwit (Limosa), speckled grey and brown, with greenish legs and a recurved beak. A large rapacious-looking bird, which came sailing majestically within gunshot, was brought down, and turned out to be the frigate bird (Tachypetes aquilus), a bird confined to tropical regions, but having a wide range throughout them, being not uncommon both in the Atlantic and Pacific Oceans. When it fell, a strong guano-like smell pervaded it, which was very disagreeable. I measured its expanse of wing, which proved to be nearly seven feet from tip to tip; and on opening its stomach I found, in a partially digested state, three large flying fishes and two squids. Small flocks of a pretty species of white egret frequently flew along the shore, and indeed, with gannets, made their appearance about the ship immediately upon her anchoring off the shoal. I shot one from the ship for examination, and found it to be 20 inches long from tip of beak to end of tail, and of a pure white colour, with the exception of a few orange feathers over the base of the beak, which formed a crest; bill yellow, and legs greenish brown. It was not provided with any of those special feathers which adorn our British species. The stomach contained a few remains of beetles.

But the dominant and characteristic bird of Pratas Island is the Gannet (Sula alba). These birds measure 4 ft. 10 in. from tip to tip of wing, and 2 ft. 9 in. total length from beak to tail, which is wedge-shaped. The head, neck, back, and tail are fuscous, breast and belly white, legs and feet yellow, and completely webbed. They are common birds on most of these islands, and are well-known to seamen. They fly heavily and usually low, fearlessly approaching within gunshot, and even stone's throw, and some of the men amused themselves with throwing lumps of coral at them as they flew by, the same bird returning again and again at the risk of being knocked down.

A walk through the interior of the island among the shrubs and bushes revealed to me the domestic economy of these birds. In the open places, and under the shelter of the bushes, the mother gannets were sitting upon their nests and eggs. The nests were mere hollows in the coral sand, strewed with a few bits of grass, with some admixture of feathers, and perhaps a piece of seaweed, forming, at best, a very rude cradle, in which were deposited two eggs. These eggs were about the size of goose eggs, white, with a suspicion of a blue tinge, not smooth and glossy like hens' eggs, but more or less scratched, as though the scratches were made when the external coat was soft, and had afterwards dried, preserving the marks. One nest alone contained four eggs. The poor bird sitting upon the nest would show symptoms of uneasiness as we approached, pecking the ground or coarse grass fiercely with its long, straight beak, but did not offer to quit the nest until we were within two or three yards of it, or even less. Then placing the end of its bill upon the ground, with a gulping effort it vomited up its meal, depositing it beside the nest, and floundering forward, took wing and rose into the air. This was the proceeding at nearly every one of the hundreds of nests which we disturbed; it was evident that the birds had just gorged themselves with food, and then sat down upon their eggs (unless, indeed, their mates had brought them food, a circumstance which I did not see myself), and that they were unable to raise themselves off the ground until they had got rid of the superfluous weight in their stomachs. On examining the vomited food, I found it to consist invariably of flying-fish, generally of a large size, and usually but slightly digested. There were sometimes six or seven of these fish, in other instances only three or four, and in two or three cases a squid or two intermixed with them. But what numbers of flying-fish must exist in the neighbourhood to afford such a daily supply to so large a number of birds; and yet we did not see a trace of flying-fishes about the island, and might otherwise have supposed there were none. Meanwhile the gannets formed a thick cloud overhead, the noise of whose screams and the rustling of whose wings formed a wild accompaniment of sounds. They flew so close overhead that we could have knocked them down with a stick in any numbers, and I was obliged to wave my gun about as I walked along, in order to keep them from carrying away my hat. By degrees the birds rose higher, and those we had disturbed returned to their nests as soon as we had passed a few yards beyond.

In the latter part of the afternoon a seining party came from the ship, and the nets being prepared, four casts were made very successfully. A great number of fish were taken and stowed away in the sail-bags, but it was too late and too dark to examine them very closely, and they were distributed amongst the ship's company and dressed for breakfast. Among them were a great many of a large silvery mullet; no flying-fish, however. In one of these hauls the net was so impeded by the quantity of the reticulated Ulva before mentioned, that it was drawn in with great difficulty.

It was now dark, and a breeze was springing up. A blue light burnt from the shore was answered by another from the ship, thus distinguishing her position, and having embarked in the gig we were soon scudding along under sail. Meantime the full moon rose grandly over the sea, and in half-an-hour we had measured the way back to the ship which it had taken two hours' hard pull to do in the morning.

The towing-net hanging out from the ship when lying off the island was, the first evening, filled with a dense brown deposit, which on examination proved to be composed solely of Zoëæ, or crab-larvæ, all of the same species. The next morning on raising it again in the same spot, not a Zoëa made its appearance, but instead of them were numbers of Leucifer, Entomostraca, and other minute Crustacea, also little Atlantæ, fronds of reticulated Ulva, and decaying leaves of Zostera, upon which were Orbitolites, Spirorbis, and minute Polyzoa.

A strong north-east wind prevented us the following day from paying another visit to the island; while, lying under its lee, we remained at anchor for the sake of the shelter it afforded us. But on the second day, towards sunset, our attention was attracted by the curious phenomenon of long rolling waves coming in from the south-west, which increased as the evening advanced, causing considerable motion in the ship. Towards midnight these south-west rollers increased to such an extent, the wind still blowing strong from the north-east, that it was deemed desirable to slip cable and put to sea, since the proximity of the reef was very undesirable if bad weather set in, while the rolling swell endangered our bumping upon the reef in a spot where our fair-weather anchorage left but little room to spare. We kept outside the edge of the reef therefore during the night, and next day approached its north-west corner. Here we saw the terrible sight of the long line of breakers on our lee side, extending for miles along the northern edge of the reef, over which the sea, lashed into foam by a strong breeze of some days' duration, was dashing wildly in a broad straight band of white foam. Finding that the wind freshened, and that we could do no more at the Pratas Shoal, we steered north-east and left the dangerous reef behind.

The explanation of the curious phenomenon of south-west rollers coming in with a north-east wind followed in due time. They were caused by a typhoon which was blowing between 200 and 300 miles to the south of us, and which *recurved* in lat. 16° 10′ N. and long. 116° 30′ E., according to the observations of Capt. Symington, whose ship, the "Northfleet," was twice caught in it, and who published an account of the Cyclone.

Pratas Island being so small a spot, and situated 170 miles from the mainland of China and about 250 from Formosa, it is remarkable that so many land-birds should have found a home there; and the incidents of the two or three days which elapsed during our passage from the reef to the Island of Formosa were particularly interesting, as throwing light upon this circumstance. Steering north-east for Takau-con, we experienced a strong head-wind the whole way, that is, the direction of the wind being in a straight line from South Formosa to Pratas Island. We left the reef on May 3rd; on the 4th a large flock of sandpipers met us, going with the wind towards Pratas, where no doubt they would find a resting-place. But the following day, being then a little more than halfway from the reef to Formosa, the rigging was scarcely free at any time during the day from feathered guests, which must have been driven off the Formosa coast by the wind, and some of them at least would have reached Pratas had they not found a resting-place, and in some instances a passage back, on board the "Serpent." The following birds I observed at various times during the day, and sometimes several of them flying about the ship, and from time to time settling on various parts of the rigging:—a yellow warbler (Sylvia); a yellow wagtail (Motacilla boarula); a shrike (Lanius), grey with a black moustache, apparently identical with the one already seen on the island; two species of swallow (Hirundo); a small heron (Ardea); a handsome black bird rather bigger than a common blackbird, with a crimson beak and a large white spot on each wing; a red dove with a white head; a yellow and black spotted plover (Charadrius pluvialis orientalis), precisely resembling the British golden plover; a species of flycatcher (Myiagra azurea); and a bird closely resembling a hen chaffinch (? Munia topela).

This interesting assemblage of birds was evidently but a few of the numbers blown off the land (probably Formosa) by the force of a moderately strong north-east wind, and of them, many would perish in the sea, a few would find relief and restoration in passing ships, and without doubt some would reach Pratas Island, and finding means of subsistence would take up their residence there, and be jotted down in the Avi-fauna of the next observer.

Entrance to the Harbour of Ta-kau.

CHAPTER III.

FORMOSA.

TA-KAU-CON, AND THE PESCADORES ISLANDS.

Character of Native Race —Dutch Occupation —Treaty Ports— East Coast— Arrive at Ta-kau—Lagoon—Apes' Hill—Land Crabs—Leaping Fishes —Walk in the Country—Water Buffaloes—Padi Birds—Village of Pi-hi-kun — Chinese Ladies — The Pescadores — Ponghou — Makung— Cheap Provisions— Cuttle Fish — Absence of Trees and Birds —The Rocks—Visit the Mandarin—Photography—Wreckers.

THE Chinese do not appear to have been acquainted with the existence of the Island of Formosa, or Tai-wan, until the year 1431 A.D., a circumstance which does not speak much for the naval enterprise of a people who had possessed the mariner's compass for so many centuries. It was originally inhabited by a race who were described as—the men of tall stature, very corpulent, and having a complexion between brown and yellow, who went naked during the summer—

without blushing, adds the Dutch chronicler: the women, of short stature, yet corpulent and strong, of a lighter complexion than the men, well dressed, and exhibiting a natural modesty. Both sexes friendly and good-natured, they would not readily cheat or steal, not treacherous like other Orientals, anxious to learn; the men, however, averse to labour, so that the women had to do all the work of the field, and the heaviest work at home. Formosa was discovered for Europe by the Portuguese, and from its pleasant aspect called by them Ilha Formoza, which name it has retained. The Dutch, however, who found the natives as above described, occupied and colonised the island, and doubtless did much good there; raised the people from a state of barbarism, educated them, and instructed them in the Christian religion. The Chinese, already conscious of the advantages of settling in so fertile a country, treated the aborigines like dogs—robbing and murdering them as it suited their convenience; and it is no matter of surprise, therefore, that the natives felt an attachment to the Dutch, who enforced their own laws, prohibited fighting among them, made the education of their children compulsory, and left them nothing of their own barbarous customs and laws, except the privilege of selecting their own chiefs to manage the affairs each of his own village; each chief being himself under the jurisdiction of a Dutch military officer, who, with 25 men, was stationed in every village of importance.

The aborigines of Formosa are reputed still to have a traditional reverence and regard for white men, and it is much to be regretted that so firm and benignant a rule as the Dutch seem here to have inaugurated should have been cut short by an overpowering attack of the neighbouring half-civilised Chinese.

Formosa is now opened up once more to western enterprise; but in a very different manner from the time when the Dutch philanthropists occupied it. It is still in the hands of the Chinese, who reserve their monopolies of some of its most important productions, such as sulphur, camphor, rice, &c. By treaty, the ports of Ta-kau in the south, and of Tam-suy and Ke-lung in the north, are open to foreign trade, and a few merchants have settled in these places. The capital of the island, however, Tai-wan-foo, being situated nearly three miles inland, up a muddy and shallow river, is very unsuited for commerce or for residence, and although our consul, Mr. Swinhoe, who has done much for the zoology of the island during his residence in it, first planted his consular flag here, he soon found it desirable to remove it to Ta-kau. But still the resources of the country are undeveloped, and it yet remains for some enterprising nation to do justice to Formosa. Chinese policy only stunts the growth of its commerce, and, dog-in-the-manger like, most imperfectly and insufficiently does that which it will not allow any one else to share in, except at a disadvantage.

The western side of Formosa only is occupied by the Chinese. The eastern rises for the most part into a range of lofty mountains, in the recesses of which still dwell the aborigines, with here and there perhaps a small community of Chinese, who are more or less in awe of their savage neighbours. This side, too, is very rarely visited by Europeans, being almost devoid of harbours, and the coast inhospitable and dangerous. The only harbour, in fact, upon the east coast is that of Sau-o bay, concerning which more will be said in another chapter.

This interesting region we were now approaching, with the

probability of spending some weeks in visiting its various ports; and on the 6th May the "Serpent" arrived off Ta-kau-con, in the south-eastern corner of the island.

The harbour is so small, and the entrance so narrow, that we did not attempt to take the ship in, but contented ourselves with anchoring outside, where heavy rollers, the result of the recent typhoon, were setting in from the south-west. Several catamarans—mere rafts of bamboos, on which a single Chinaman stands and rows—came off with vegetables and fruit, presenting a curious appearance, for not only were they entirely lost to sight when in the trough of the sea, but even when borne up on the crest of the wave the rower seemed to be standing upon the water itself.

The aspect of Ta-kau from the anchorage was striking and interesting. North of the harbour was Apes' Hill, consisting of a double truncated elevation, the higher plateau reaching 1120 feet—and, southwards, the Saracen cliffs, a long line of low perpendicular rocks, upon which a few cycads were growing. Between these elevations was the narrow entrance to the harbour, within which could be seen the yards of several square-rigged vessels mounting Bremen colours, while behind all was a magnificent range of mountains in the distance—a portion of that chain which traverses the island of Formosa from north to south—whose slopes and base are the abodes of numberless species of deer, wild cats, pheasants, &c., and which formerly had the reputation of harbouring tigers also. But we have much to learn yet of the natural productions of the island; and but few Europeans have penetrated even to the foot of these hills, about which we know but little more now than we did when the forgeries of Psalmanazar gulled a susceptible public.

On rowing into the harbour, the numerous picturesque junks anchored within gave it a foreign appearance, very striking to one who, like myself, now entered a Chinese port for the first time. On either side houses, including some in European style, were scattered—the real Chinese town forming a long, narrow, dirty street, similar in character to those which I shall have occasion to describe in other parts of Formosa. It is situated directly on the shores of the harbour immediately on entering, and is inhabited by a very low and poor coolie class of Chinese. The European community at Ta-kau is very small, consisting of a vice-consul, one or two English merchants, two medical gentlemen—one of whom, Dr. Maxwell, is a medical missionary—and a commissioner of the Imperial customs.

The harbour opens into an extensive lagoon which runs some miles inland, and is separated from the sea by a narrow strip of slightly-elevated land, which serves as a mole. From the hills in the neighbourhood of the harbour this lagoon may be seen stretching away through mangrove-covered flats, among which boats could be seen threading their way. Beyond this, a wide and fertile plain of alluvial soil, covered with padi fields and other cultivation, swept up to the base of the magnificent mountains already mentioned, and was dotted with villages, clumps of trees, and other elements of a luxuriant landscape; while out to seaward the small island of Lambay broke the monotony of the view in that direction.

Apes' Hill is so called from the fact that a (tailed) species of monkey is occasionally seen upon certain parts of it; but as far as I could learn, they are difficult to meet with, though I was assured that they really existed. I ascended to the summit, which was very rugged, the side next the sea being rocky and precipitous; and as it was this part which the monkeys were said chiefly to inhabit, I did my best to get a sight of them. Lying flat down, therefore, I looked over the edge, but neither the dislodgment of stones nor the clapping of my hands succeeded in eliciting any traces of the animals, which, in fact, appear to be almost as mythical and rarely seen as the true apes on the rock of Gibraltar. While thus engaged, a loud rush near my head made me retreat from my insecure position, and on looking up I found that a number of large kites (Milvus govinda), which were always hovering about the coast in search of garbage, had assembled overhead, and one of them had made a swoop near me, probably to reconnoitre the unusual object.

The lower part of Apes' Hill consists of rugged coral blocks, embedded among which I obtained a few recent shells. The blocks are thrown up in a very loose manner, but for the most part covered with bushes and herbage, even up to the summit. Abundance of a species of Euphorbia, and stunted bushes of guava (Psidium) grew upon the sloping sides, while near the summit appeared the characteristic cycads, which were now in flower, and might easily have been mistaken at a distance for small palms. Among them flew in considerable numbers a large, red-winged orthopterous insect (Gryllus), and at the summit was a small green species, with the head singularly elongated and produced in front, belonging to the genus Tryxalis, which seems largely represented in the island.

Upon the shores of the lagoon was an excellent spot for watching the habits of the land crabs (Gelasimi), which marched about in a serio-comic manner amid their holes; each one as it cautiously moved along held up in front of its eyes its single large and delicately-tinted claw, with an expression half of defiance, half of defence. Prowling thus about, probably in search of food, they were readily alarmed, and retired to their holes, which generally seemed too small for them, so that it took a little time for them to accommodate themselves to their narrow dimensions. If closely pursued, therefore, they were easily captured. I carried one to some distance, and placed it at the mouth of another hole, down which it immediately dived and disappeared, and although I waited a considerable time in the expectation that the tenant of the hole would drive it out and show some displeasure at the intrusion, nothing of the kind occurred.

Another singular animal which I saw here for the first time, but which I found numerous on many subsequent occasions, was the leaping fish, Boleophthalmus Boddaertii. These curious salamandrine-looking creatures, for it was difficult at first to say which they were, contrived to elude pursuit in the most active and provoking manner. Each step in advance caused them to jump, jump, in a rapid and agile manner from almost under my feet—for when at rest they were scarcely distinguishable from the mud on which they were lying, and to which they admirably assimilated in colour—but on the least alarm they would make a series of leaps, which rapidly brought them down to the margin of the water, and from which it was next to impossible to cut them off. They are wedge-shaped in form, usually about 3 or 4 in. long, with flat pointed tails and broad heads, upon which is situated a pair of prominent eyes. They have been called by sailors "Jumping Johnnies," and are by no means confined to muddy or sandy shores, for I have found them equally among smooth rocky places, up which they climb with great skill, by a series of leaps, wriggling and curving the tail at each leap in a contrary direction, that is, to right and left alternately. Their leaps are effected by means of their curiously-bent ventral fins, which look something like a pair of hands placed immediately behind the head, and as they always make straight for the water and *double* with great agility, it is scarcely possible to capture them excepting with a net.

The vicinity of Ta-kau is fertile and highly cultivated, and the country populous and interesting. The lagoon has the appearance of a broad river, with mangrove-bordered creeks and numerous large arms, and at its head is a muddy expanse, given up to hosts of land crabs (Gelasimi), whose myriad holes give forth a crackling sound as their tenants withdraw themselves on the approach of footsteps. Beyond this, padi fields cover the greater part of the country, among which numerous villages stand like habitable spots of *terra firma* amidst a marsh. The padi fields are for the most part rectangular, with narrow ridges between them, which afford a precarious footing, and render it necessary to keep a careful eye upon one's footsteps; for the rice grows up from pools of muddy water, into which an indiscreet step would at once plunge the incautious pedestrian. Upon these waters, numerous aquatic insects (Hydrometræ and Gyrini), of species indistinguishable at first sight from those in English ponds, were sporting; and many large shells, chiefly Paludinæ (P. æthiops and P. chinensis), were floating among the stalks of the rice. Strange as it may appear, the aspect of the scene forcibly reminded one of English cornfields in spring—the green rice hiding the unsightly marshy aspect of the country.

Nestling amid the trees, among which bananas and bamboos held a conspicuous place, were numerous villages, the houses of which were usually plastered over with mud. Beside them were small cottage gardens, and plantations of sweet potato (Convolvulus batatas). From their villages groups of curious natives came out to see us; noisy dogs rushed out, barking, and ran away growling; and great hollow-backed pigs, of the real Chinese breed, grunted lazily from the mire in which they were wallowing; while here and there, in a secluded spot, was tethered a water buffalo (Bos bubalus), one of those unsightly brutes which represent the domestic cattle of China, his black hide plastered with mud half dried, and his neck stretched out with a stupid and frightened expression. No sooner did we appear in sight than, in many instances, the animal, clever enough to recognise strangers, began to caper about, and, violently snapping the cord which was fastened to a ring through his nose, went crashing through the bamboo fences into the plantations, with the effect of quickly bringing out his wrathful master in hot pursuit. In some spots we came to a herd of these animals bathing. They delight in water, and in wallowing where the mud is deepest and softest; and they require no persuasion to go into a pond, however thick and dirty, but, laying themselves down, they remain with their noses just above the water for any length of time. Such herds were usually under the charge of two or three lads; and the animals, on seeing us approach, immediately began to stretch out their necks, regarding us with a stupidly vicious stare, as though they would immediately quit the water and rush at us. The former they would probably have done had we not been very circumspect, and their guardians were in great fear of their rushing out and being dispersed; but there was little chance of their running at us, for they would more probably have stampeded in the opposite direction.

The most common bird was undoubtedly the Padi bird, a species of heron (Ardea prasinosceles), which was constantly flying over the padi, or rice-fields; and it was also accompanied by a pretty white egret (Herodias garzetta); but on the banks of a small lake a cluster of trees was full of these birds, whose colours were relieved by two other species, one (Buphus coromandus), which possessed a number of rich buff feathers; while the other was of the ash grey of our ordinary heron (Ardea cinerea), which is here common. Vast numbers of these birds, all mingled together in the trees, were set off by the thick green foliage, and had a very pretty effect. They kept up a loud and constant chatter, and seemed all disputing with one another for the possession of nesting-places. As we returned to Ta-kau we captured a splendid night heron (Nycticorax griseus), a truly nocturnal bird, but the exigencies of whose young family required it to be abroad in the day at this season; and among the smaller birds, the most notable were two species of flycatcher—one, Myiagra azurea; and the other, Ixos Sinensis. Several pretty doves nestled up in the trees, among which I noticed Turtur humilis by its peculiar coo; and on the lagoon a summer-snipe (Totanus) afforded practice for our guns.

At the village of Pi-hi-kun we halted to refresh, and were soon surrounded by an admiring group of villagers, who turned out to gaze at us, and crowded round with the greatest curiosity to see the foreigners eat, and to examine all their accoutrements. The gun, powder-flask, and shot-case came in for their share of admiration, which was at its height when we brought down a Padi bird as it flew over the village. Our clothes, their texture and cut, were curiously inspected, and all the contents of our pockets were turned out, the old men being as inquisitive as the youngest, but all civil and good-humoured. It was my telescope, however, which caused the greatest *furore*, and all in turn were treated to a peep through it. Not in the least degree backward was the irrepressible boy, who, in Formosa as everywhere else, maintained his character for impudence and inquisitiveness. We became very popular, and water was brought us in a gourd, and pine-apples produced, which assisted in extinguishing hunger and thirst at the same time; and when at length we left the place we were escorted out of the village by a crowd of *gamins*, to whom the day's excitement was something to be talked about for a long time after. The girls and young women, however, were timid and backward, sometimes venturing into the skirts of the crowd to get a stolen look at us, but immediately retreating to a safe distance if they saw that they were observed.

The women of the better class in this part of Formosa dress in the most brilliant colours, and numerous parties which we met walking out in the cool of the evening were amusing impersonations of the Chinese pictures and figures long familiar to us. The ladies, of whom, with children, these parties usually consisted, were, like all the females of Formosa, small-footed, and supported their difficult and tottering steps with a long walking-stick. Their dresses, consisting of a wide-sleeved tunic, cut in the formal style universal among Chinese ladies, were of the brightest scarlet, blue, or orange, embroidered with black, which contrasted well with the colour; and their full trousers were of some other equally showy material. In their hair, dressed in the elaborate Chinese tea-pot fashion, they wore artificial flowers made of the pith of the rice-paper plant, of Amoy manufacture; and as they walked painfully along, with the hobbling gait peculiar to their hoof-like feet, their figures swaying to and fro, and their arms more or less outstretched to balance themselves, they had, to us, a most grotesque appearance—but in Chinese eyes the acme of grace and loveliness, which they figuratively liken to the waving of willows agitated by the breeze.

After three or four days' stay at Ta-kau-con, we steered towards the Pescadores islands, a group between Formosa and the mainland, sometimes called the Ponghou Archipelago. This cluster consists of 21 inhabited islands besides several uninhabited rocks, lying between the parallels of 23° and 24° N., and are included with Formosa in the Chinese province of Fo-kien. A strong breeze kept us rolling tremendously as we crossed the channel, and it was a matter of congratulation to have reached the outlying rocks of Three Island and Round Island, and to get under the lee of Ponghou, the principal island of the group, along which we coasted more quietly. This gave me, moreover, an opportunity of examining the remarkable structure of the neighbouring members of this group, which all presented a peculiar flat and truncated appearance. This was particularly observable in Table and Tablet Islands, both of which consisted of flat tables, about 200 feet high, supported above upon well marked basaltic columns, and sloping from these down to the water's edge, just as is seen on the Antrim coast. So also the large island of Ponghou exhibited a columnar structure in several places, often with a sandy beach at its base; and on approaching Pong Point, the south-western promontory, I observed the columns to be broken off close down upon the beach, forming a causeway in two places, though on a smaller scale than at the Giant's Causeway. The absence of trees from all the islands gives them a rather dreary aspect. We entered Ponghou harbour and anchored near the town of Makung, the chief town of the archipelago, and were immediately saluted with half-a-dozen guns fired by some junks lying farther in the bay, though with what object we could not tell. Two or three boats presently came alongside, with persons of a very civilized and decent appearance, and by no means the wild-looking and half-clad fellows who might have been expected to inhabit such a remote place. We found it difficult to communicate with them, however, for although a race of Chinese, our China boys could not readily understand their dialect, nor could they make themselves understood. We landed shortly after at a large old Dutch fort, which once commanded the harbour, and in which a number of rusty guns were still lying in the ruined embrasures. The beach was strewn with numerous worn blocks of coral, and several fishermen were living under their boats, which they turn up at night, to shelter them against the wind.

We were very soon surrounded by an admiring crowd, composed principally of the irrepressible boys, for although some men followed us with them, no women were seen. The men and boys usually wore blue turbans, and the women, when we saw them, had universally small bandaged feet, and wore bunches of artificial flowers in their hair, as we had observed them to do at Ta-kau—ornaments imported from the opposite city of Amoy. The people generally struck us as being decently clothed, and presented a marked contrast to the squalor and dirt everywhere visible among those we had hitherto seen in Formosa. The boys also were usually neatly dressed, and there was a something in their behaviour which gave an impression of good breeding, such as we were surprised to meet with in this isolated region. We entered a boys' school at the outskirts of the town, where every one, from the schoolmaster to the smallest boys, seemed to enjoy the novelty of the visit, and to wish to show us attention. They exhibited their books, and, for a few cents, even willingly sold some of them, in which the youngsters had been drawing heroes and idols, in all the grotesque attitudes in which the Chinese appear to delight. Followed by an attendant crowd, we walked through the streets of the town, which were usually narrow, and covered over with a screen of rattans or bamboos, which formed an effectual shelter from the sun's direct rays, and kept the street cool, as is the fashion at Suez. The shops were spacious and cleanly, and the articles exposed for sale very various, but all of Chinese manufacture, and chiefly from Amoy. No European goods were visible; indeed the only article of foreign make which we encountered was some red serge. The houses are nearly all built of blocks of coral cemented together, and the tiled roofs are peculiarly curved in the characteristic Chinese manner. In the outskirts we occasionally saw women and children sitting at the doors; but as soon as they caught sight of us at the end of the street, they would hastily jump up and rush alarmedly in doors, and bar themselves in—though sometimes curiosity seemed to get the better of their timidity, and they might be seen peeping at us from behind their grass screens. If a girl ventured into the skirts of the crowd which surrounded us, a look was sufficient to drive her away; the moment our eyes met, she would sidle off confusedly, and get out of sight; children scampered away screaming whenever we appeared; and the dogs invariably singled us out, barked sullenly, and ran off to a safe distance—their exit being much hastened by the sight of a stick, for they are the most cowardly of brutes, and in this particular town often frightfully mangy and wretched-looking, much more fit to be shot than to be wandering about the streets.

Provisions were exceedingly cheap at Makung. When Her Majesty's ship "Swallow" visited the harbour recently, eggs were purchased at the rate of 300 for the dollar, and a calf cost but one dollar. When a foraging party from our ship went ashore, they purchased a calf for two dollars, and eggs at one dollar the 150, and other things in proportion. A large basket of the ground-nuts (Arachis hypogæa), a very favourite article of food in China, all ready husked, cost only 60 cents, and four dollars the picul (133⅓ lbs.) were asked for the very best rice.

Beyond the town, the harbour terminates in a broad, extensive, shallow bay, which at low water affords employment to a large number of people, who wade over it in search of shell-fish and other articles, which they consume largely in their diet. Women are principally employed in this business, both here and elsewhere, and they carry with them a basket, and a little iron hammer and pick, with which they pull out the animal from the narrowest crevices of the rocks. In some parts of the town, large heaps of shells belonging to the subgenus Modulus were to be seen, forming incipient kitchen-middens, and illustrating at once the chief molluscs of the bay, and an article of considerable consumption by the people. Haliotides (sea-ears) are also sold in the market place, as well as cuttle-fish, both fresh and dried, all of which enter into their dietary. We obtained one of these large cuttles, or more properly calamaries (Loligo), with the intention of trying its esculent qualities; but whether the fault of the cooking or otherwise, even though curried, we did not care to repeat the trial. When quite fresh, the large maculæ, and fine spots on the

E

surface, were in a constant state of change, the colour coming and going, from alternate contraction and expansion of the pigment vesicles, without any direct irritation. When pale, the colour could be made to re-appear by drawing the finger along the skin, but the power of contraction appeared to be lost when the vesicles had been cut through. As it lay on the table during the night, I cast my eye upon it, and observed that it was luminous—a glow of whitish light irregularly illuminating its whole surface. At this time it was quite dry, and the luminous appearance was not altered by passing my finger over it.

On enquiring for shells, a good many were by degrees brought to us, chiefly consisting of common cowries and harps, and olives of several large and handsome varieties of Oliva erythrostoma; but nothing else could we obtain here, though, if we had remained longer, it would perhaps have been possible to have procured others.

All these islands appear to be very destitute of trees; and standing on the high ground of Observation Island, on the opposite side of the harbour, I looked in every direction for a tree or bush, in vain. Although, however, the volcanic structure of the island is not favourable to the growth of wood, many very pretty flowers abound, the commonest of which is a species of Cassia. Probably on account of the deficiency of wood, very few birds were to be seen. A few terns flew about the harbour, and some summer snipes (Totanus) were seen occasionally. The commonest bird was the tree-sparrow (Passer montanus), abundant everywhere in the East, where it takes the place of the common sparrow of Great Britain (P. domesticus); and besides these, I observed a small shrike, and a number of larks (Alauda cœlivox) upon Observation Island, whose habit and character of song were precisely similar to those of the skylark of our own country.

The rocks of Makung Harbour, which I had an opportunity of examining, were basaltic in formation, washed smooth by the waves, and in some spots exhibiting in section the columnar structure. No seaweeds grew on these rocks, with the sole exception of the peacock-tail (Padina), which was abundant, nor could I meet with any echinoderms (starfishes, &c.). Indeed, the coast was extremely barren, and produced little else than small Paguri (or hermits) in shells of Murex, Litorina, &c., small Chitons and Patellæ. Ligiæ ran over the rocks, gleaming with rich metallic blue, and darkening them in crowds, here as nearly everywhere; and I really believe that these are the most abundant of all crustaceans, at all events of those seen. The only animal of interest I met with was a very handsome Doris, of a deep blue colour, spotted with yellow, and with branchiæ and tentacles of a bright vermilion. This richly-coloured species may be the Doris Barnardi of Kelaart (MS.). Under the stones were numerous small porcelain crabs (P. platycheles). An attempt to dredge in the bay was only rewarded with bags of mud containing a few broken bivalves.

Before quitting Makung, we paid a visit to the chief Mandarin of the place, but were not successful in seeing him at his *yá-mun*. The appearance of a foreign man-of-war in the harbour was embarrassing to the official mind, and from its rarity was somewhat alarming, inasmuch as the poor Mandarin probably was unable to conceive of such a circumstance without accompanying demands, or that it could possibly happen without any further reference to him than a mere polite visit of ceremony. He had, therefore, given out that he had gone to Ta-kau, with which answer we had, of course, to be satisfied; but the lad who guided us to his house had probably a pretty correct appreciation of the *situation*, when he grinningly hinted, "Mandoli too muchee fear."

After three days' stay we quitted Makung, our chief engineer, Mr. Sutton, an excellent photographer, having taken some views in the town on the morning of our departure. On this occasion the crowd was with difficulty kept off from the apparatus, their extreme curiosity proving rather inconvenient. One man, while a picture was being developed, and attention temporarily withdrawn, furtively drank the contents of the bottle of glacial acetic acid, and it was well for him it was not something even more deleterious. Another, who was more impudent than most of his neighbours, accepted the challenge to be painted with the nitrate of silver solution. Accordingly he received a moustache, beard, rings round his eyes, &c., which were beginning to darken in the sunlight as we left the scene, greatly amused at the surprise which awaited our forward friend when the full effects of the solution should become developed; but, unfortunately, we had no opportunity of seeing him in his altered aspect, though we may imagine it would be a source of no small embarrassment to him, and amusement to his pitiless neighbours.

A few months subsequent to our visit to the Pescadores, two English ships were wrecked in the neighbourhood. The first of these, as soon as she was observed to be upon a reef, was surrounded by 30 boats, and some 300 natives boarded her and looted the ship of every movable article. They do not appear to have offered any personal molestation to the Europeans, who were even accommodated with the shelter of a joss-house; but their goods were taken as something which had fallen to the plunderers by right. In the second case also, the European crew were stripped and robbed; but, otherwise, the intruders showed an inclination, provided good remuneration were offered, to assist the captain out of his difficulties. Not everywhere on the Chinese coast is so much forbearance shown as by these islanders.

CHAPTER IV.

FORMOSA (*continued*)—TAM-SUY.

Towing Net in Formosa Channel—Pterosoma—Firola—Sagitta—Atlanta—Glaucus—Alima—Phyllosoma, or glass-crab—Cerapus—Hyalæa—West Coast of Formosa—Fort Zeelandia—Notonectæ—Arrive at Tam-suy—The Harbour—Boulder Clay—Chinese Graves—Rice-paper Plant—Bamboo—The Town—People—Rice Embargo—Visit to Mbangka—Camphor Monopoly—Visit the Chief Mandarin—Return Visit—Queen's Birthday.

At daybreak on May 15 we weighed and stood out of the harbour of Makung, first directing our course towards a supposed shoal, marked doubtful on the chart, which we did not, however, succeed in discovering. But the appearances were quite sufficient to deceive the inexperienced—such as long lines of ripple caused by the rapid north and south tide of the channel, and drift dust in the distance looking like breakers. The mast-head man also reported shoal water; but it proved to be a fallacious appearance caused by the tide rips, which ran so strong that the towing-net could not be kept out except at slack water.

And here I may refer to several singular marine animals, discovered by the towing-net in the Formosa channel, which proved a rich locality for strange and rare forms. Among them was the Pterosoma (Pt. plana), a transparent, delicately-tinted winged animal, thick and gelatinous, and almost invisible in the water. It belongs to a class of mollusks known to naturalists as Heteropods, oceanic animals of anomalous forms, with the foot variously modified for swimming. The Pterosoma was established as a genus by Lesson, upon a species he found swimming in the vicinity of New Guinea; but either the drawings of his animal are very badly executed in all the books, or the one found in my net must be a second species, for there is but little resemblance between them. Another delicate animal of the same class was the Firola, a transparent creature, with a long proboscis, and swimming by means of a well-developed fin in the lower part of its body. A third was still more curious—an elongated, transparent body, without eyes or tentacles, but furnished with two pairs of fins and a fish-like tail, the whole body like a minute arrow, and hence called Sagitta. It darts through the water by sudden and instantaneous jerks, during which it is lost to view for a moment. So transparent is the body that the whole internal organisation may easily be observed, and the circulation of granules, upwards (towards the head), in the neighbourhood of the tail on either side the body, and in the middle downwards towards the tail. This animal is referred by Prof. Huxley to the articulate division of animals. Another of these nucleo-branchs, as they are termed, because their respiratory and digestive organs form a kind of nucleus on the posterior part of the back, was the pretty little curly-shelled Atlanta —shell and animal equally transparent, the latter with eyes and tentacles, and moving actively by means of a fan-shaped fin. All these delicate oceanic animals have a remarkable range, being found for the most part both in the Atlantic and Indian Oceans, as well as in the Mediterranean Sea.

Only once did I meet with the little purple Glaucus, an oceanic nudibranch, of which so much has been written. This sea-lizard, as it has been called, soft and fragile as it is, is a very tyrant over animals beautiful and delicate as itself, and the pretty blue Porpitæ are the victims. But as this was rare, so the glass-like crustacean, Alima hyalina, was common in the net—lovely forms, whose carapace seemed carved from the purest crystal, with an elegance of sculpturing and sharpness of outline that could not be surpassed,—perishable animals, but which, while they remained alive, were active in the water. The only spots of colour in their bodies were their two eyes, mounted on long stalks, and giving out a rich golden-green glow, which was positively luminous. Almost equally transparent were the glass-crabs (Phyllosoma), whose flat, leaf-like bodies and long branched legs seemed as though made of fine plates of clear mica. These nocturnal oceanic animals (for they never appear in the net by day) are, however, very passive and quiet, and seldom show any signs of life. It is not improbable that they are larval forms of some possibly altogether different beings.

To find caddis-worms in the towing-net seemed remarkable; but small worm-like crustacea (Cerapus), furnished with large antennæ, and living in tubes or cases, were not unfrequently met with. These little creatures have usually their head and foremost legs peeping out of their case, which seems to be just large enough for the body; but alarm it, and it vanishes within, re-appearing immediately, head first, at the other side, so that one can hardly be persuaded that it has not two heads. Although some were minute, others were of considerable size, and much larger than those usually described. More than once, Hyalæas, and other graceful Pteropods, were captured; but one of these, probably the Hyalæa tridentata of Lamarck, presented appearances such as I have nowhere seen described. When first taken, the keeled lower angles of the globular shell showed nothing worthy of remark, the appendages were small and contracted; but gradually they became spread out to their full size, and became large, oval, semi-transparent leaves of a light green colour, exceeding in length all the rest of the body, now hanging straight down, and now more divergent. The animal could contract them at pleasure, and in a moment spread them out as before. The shell itself was tinted with rich brown; and it appeared to have three pairs of fins, the largest and uppermost brown, a smaller pair of a reddish tinge, and a third pair transparent and projecting somewhat backward over the convex side of the shell. The edges of the wings (or fins) and the points of union of the green leaf-like expansions to the other parts of the body, were of so delicate a structure as to be invisible, except on close and careful inspection. This animal swam rapidly in a horizontal direction, and kept itself floating on the surface of the water by a butterfly-like movement of the fins; but when at rest, it kept them folded over the convex side of the shell.

The whole west coast of Formosa, between Ta-kau in the south-west and Tam-suy in the north, is very flat, consisting for the most part of low alluvial plains, with no conspicuous elevations. The mountain range which culminates in Mount Morrison, and renders the east coast harbourless by its near approach to the sea, nowhere comes near the western side. As we advance northward from Apes' Hill, the coast becomes low and level; little flat islets appear at intervals, which are seen to be connected by sand-banks on a nearer approach. The capital, Tai-wan-fu, not many miles above Ta-kau, is invisible from the sea, being situated some two or three miles up a muddy river: at its port, however, 16 large junks and a square-rigged Bremen vessel, as well as numerous fishing-boats lay at anchor. Near the mouth of this river also are the ruins of an old Dutch fort on the beach, celebrated in the annals of the island as Fort Zeelandia, and more particularly in connexion with the tragical episode which ended the Dutch occupation in 1661. Formosa, under its enterprising colonists, had reached a political and social condition far superior to that which it now enjoys, and an attachment had sprung up between the natives and their foreign rulers; but this very prosperity excited the cupidity of Kok-singa, a renowned piratical chief, who, in May, 1661, appeared with a fleet and force of 25,000 men. The Dutch concentrated themselves in Fort Zeelandia, while hundreds of the settlers fell victims to the cruel invader, whose descent was sudden and unexpected. Finding that the besieged were determined to hold out to the last extremity, the pirate became exasperated and would listen to no terms; meanwhile massacring with cruel tortures hundreds of Dutch prisoners who had fallen into his hands, after which the corpses were stripped and buried in heaps— the women being distributed among the officers and men of his force. The little garrison at length was compelled to capitulate, and the Dutch were for ever expelled from the island; while the natives, who were in a fair way of being civilised and Christianised, have, meanwhile, relapsed into their primitive barbarism. The devotion of the Rev. Mr. Hambroek, a minister of the Dutch reformed church, who was sent by Kok-singa to make terms with the besieged, is still on record—a devotion worthy of a Regulus, and bearing a close analogy to that old tale of Carthage.

North of Zeelandia is Kok-si-kon, formerly a port, but now closed up; and beyond this a long, low, sandy beach, upon which people could be seen walking, or sometimes sitting in groups to watch us; nets hanging up, with here and there a long, low hut; and after dark, a number of lights, having all the appearance of a row of gas lamps. Along all this low coast a singular aërial misty effect was observable, which appeared to arise from a lagoon behind the sandy beach. Everything seemed enlarged; men, passing by, seemed "as trees walking;" little villages appeared like large towns of stone houses, until we approached nearer, when they dwindled down to mere collections of huts. It was a kind of mirage arising from irregular refraction. In Gilim Bay 30 junks lay at anchor.

The only place where hills approach this coast is in lat. 24° 15′, where long sloping shores, highly cultivated, thickly populated, and dotted with numerous villages, skirt the ranges of high hills rising about two or three miles inland, which are often intersected by horizontal valleys of denudation, affording long and pretty vistas; the lofty mountains of the Morrison range affording a picturesque background to the whole. As we were passing this part of the coast towards evening, the cabin table became covered with small water-boatmen (Notonectæ, of the restricted genus Corixa), freshwater insects, which must either have made an unwonted flight out to sea, or have been washed off the land by the embouchure of some river which here joined the channel. They flew about the cabin and round the lamp like moths, and having placed some in a basin of salt water which happened to be upon the table, they swam merrily; but they were all dead in the morning. It had been a beautiful calm day, but in the evening a breeze sprung up along the coast.

At early morning on May 18th we were off the harbour of

Tam-suy, which, like that of Ta-kau, is well pointed out by natural landmarks on either side. Two lofty and picturesque hills render it very conspicuous, that on the north called Tai-tun, which forms an imposing ridge, rising to the height of 2,800 feet; and that on the south side, known as Kwan-yin, and having two prominent peaks, attaining an elevation of 1,720 and 1,240 feet respectively. We were soon boarded by a Chinese pilot, who was anxious to take us in, and who magnified the dangers of crossing the bar without his assistance; but our captain knowing something of Chinese character, was quite aware that the pilot was as likely to run us ashore as not, and preferred trusting to his own experience and skill. It being low-water we rowed into the harbour, reconnoitring the bar as we crossed, and proving its practicability; but, meanwhile, a breeze sprung up, and a heavy sea broke over it, while a thick haze obscured the ship and shut it out of view. An attempt to recross the bar in the boat proved unsuccessful, and we were fain to remain on shore, while the ship was forced to put to sea and stand out till morning. Our rockets and blue lights that night were unanswered, and we were therefore glad to see with the morning light our vessel once more in the offing; we speedily rejoined her, and at high-tide crossed the bar and entered the harbour, much to the confusion of the pilot, who soon after came on board and offered his services in the new character of *compradore*.

The town of Tam-suy, or as it appears to be otherwise called, Hoo-wei, is situated upon the right bank of the harbour. From land to land at the entrance is just half a mile, but a considerable spit of sand diminishes it by more than one half. Within the harbour, however, it rapidly increases to three-quarters of a mile, and even a mile in width, affording good anchorage for large vessels. Immediately upon the left-hand on entering, there is a small Chinese fort; and half a mile higher are the ruins of an old Dutch casemate—a square, red-brick building, once no doubt of considerable strength, and elevated 50 or 60 ft. above the water's edge.

This elevated right bank, upon which the town stands, presents very remarkable features. It rises in an undulating manner for about 100 ft., and is entirely composed of alluvial clay, containing a vast number of boulders of stone. These boulders are of the most various sizes, from such as can be easily lifted by the hand, to large blocks of 20 ft. in circumference. They are also of very varied forms—some being round and smooth, and evidently more or less rolled; while others are quite angular, and have little or no appearance of having been water-worn. I carefully examined many of these blocks to see if I could discover any traces of striation which could be attributed to glacial action, but although I met with some suspicious markings, I could not satisfy myself that they were scored by the agency of ice. Moreover, there was no marked difference in the various boulders as to their lithological character, but to all appearance they were, with little exception, formed of the ordinary pebble green-stone.

This alluvial soil is very fertile, and the undulatory character of the ground gives considerable picturesqueness to the neighbourhood of Tam-suy. Houses are scattered about on the hill-sides, and a large amphitheatre just outside the town forms a spacious and well-filled burial-ground, consisting of an immense assemblage of the characteristic forms of Chinese graves. These are mostly of the horse-shoe form, or rather omega-shaped, and vary in elaborate and complicated structure according to the position of the occupant. The ordinary merchant has a simple tomb, with a rectangular stone tablet in the centre, inscribed with Chinese characters in red and black; while the tombs of the Mandarins are often extensive structures, in which the limbs of the omega are enlarged into fantastic and elaborate copings of stone, ornamented with statues and carvings. The poor are satisfied with a simple mound and small sculptured headstone, or even less; though such is the veneration for ancestry, that the poorest usually find means to secure some memorial of their deceased parents.

Upon these hills grows in considerable abundance the Rice-paper plant (Aralia papyrifera); and from this place it is largely exported to China for the purpose of making upon the prepared paper those brilliant colourings for which the Chinese are so renowned. It is a small but handsome plant, the stem growing to the height of from 4 to 6 feet, and then giving off by long footstalks a number of handsome large digitated leaves of a dark green colour, but whitish beneath, which spread out sometimes 4 or 5 feet on either side. For a long time the source of rice-paper was a mystery, and its name indicates the common fallacy as to its origin; but an examination with the microscope could not fail to detect the large cellular substance of which it is really composed, namely, the little-altered pith of a plant. This pith is of a snowy whiteness, and occupies the whole of the cylindrical stem, more particularly at its upper portion, becoming smaller near the base. I never found any hollow centre in the pith, although it is said the Chinese themselves call it the *Tung-tsau*, or hollow plant; nor did I observe any specimens in the neighbourhood of Tam-suy more than 6 feet high, although the Chinese accounts make it twice that height. Probably the specimens which came under my notice were young, or those which had not had the benefit of cultivation, for they were scattered sporadically upon the hill-sides. The mode of preparing the paper from this plant is by skilfully paring the previously-removed pith with a broad and sharp knife, which shaves it cleanly off in a spiral manner from the circumference to the centre, at the same time preserving an equable thickness throughout. The substance is then flattened out, cut into smooth sheets, and is ready for the reception of pigment, which can be laid on with remarkable facility and brilliancy.

But perhaps the most prominent feature of the vegetation of Tam-suy and its neighbourhood is the bamboo (Bambusa arundinacea), everywhere a striking object from its graceful feathery foliage. It lines the river's banks, forms hedges and fences, and is remarkably beautiful. At the same time it is the most useful of trees, from which almost every article and utensil is made; the small canes, and the large heavy stems alike, with little preparation, being converted into innumerable useful objects; while the split wood is utilized in a hundred ingenious ways, and there is scarcely any manufactured article into which the bamboo in some form does not enter.

The long rambling town of Tam-suy consists, for the most part, of a narrow street of shops of a poor description, paved with great cobble-stones, or else not at all, and in which pigs of all sizes, and barking dogs, dispute the passage, which, in some parts, scarcely admits of two passengers passing one another. The Vice-Consul, Mr. Gregory, resides here, as well as three or four other Europeans, either engaged in mercantile affairs, or employed in the Chinese customs. The consulate, however, is but a poor building for the representative of Great Britain; for the inhabitants, who are mostly of the Coolie class, and upon occasion can show themselves a turbulent set, have a prejudice, forsooth, against building houses more than one story high, and no such dwelling exists in Tam-suy.

Squalid, however, and unsightly as are the buildings of Tam-suy, there is a very pretentious joss-house or temple, in which the stone pillars, elaborately carved, represent, with considerable cleverness, fantastic dragons encircling the columns in high relief, and holding loose stone balls in their mouths. Workmen were still engaged upon these sculptures.

The people of Tam-suy are poor and meanly clad, and the same may be said of the other towns in this part of Formosa. The males usually wear nothing more than a short pair of drawers, or some substitute for them, many of the younger male children going entirely naked. The women and girls, however, are always decently clothed, very few of the female children being bare even to the waist. Bandaged or small feet are universal among them, the only exceptions being a few among the lowest of the low.

Bullocks, goats, and poultry are difficult to obtain, but pigs are abundant, though few who had an opportunity of witnessing their disgusting habits and foul feeding would care to eat them. Ducks also are plentiful.

Rice is abundantly produced in the neighbourhood, as well as in other parts of Formosa, but its exportation is forbidden on pretence that no more is produced than is required for home consumption. This embargo was issued by the Tao-tai of Tai-wan in 1864; but inasmuch as the approbation of the foreign ministers of Pekin had not been previously obtained, it appears to have been illegal. Moreover, the Chinese authorities winked at the exportation by natives, and junks laden with it left Tai-wan in spite of the embargo, greatly, no doubt, to the advantage of the mandarins. The excuse that no more was produced than was required was simply a subterfuge; and the evil effects to the commerce of the island are evident from the fact, that it was roughly estimated that the direct loss with regard to Ta-kau alone, in commissions, was equivalent to 63,000 dollars per annum as long as the prohibition lasted. Although, however, it extended to all Formosa, it was enforced with far less stringency at Tam-suy than in the other ports.

A Hamburg merchant, Mr. Millisch, residing at Mbang-ka, or Bang-ka, situated nine or ten miles up the right branch of the Tam-suy river, having invited us to visit him there, we took the opportunity of seeing a town which, being the chief of the Hoo-wei district, was more considerable and interesting than Tam-suy. Mr. Millisch was the only European resident there, and occupied a handsome two-storied house, the only one I observed in this part of Formosa. We accordingly went up with the tide in the captain's gig, aided by a breeze from the sea. For the first four miles the stream was of varying width, averaging about a mile, and running in a south-easterly direction at the foot of the Kwang-yin hills, which, seen in the light of a western sun, had a remarkably piled-up or cone-in-cone appearance, and at the base are perforated with caverns. On the right bank a cultivated plain stretched to the foot of the Tai-tun hills, which expanded to the eastward as we proceeded. At length, at a village called Kan-tow, the stream divided, the left branch continuing its course across the island in an easterly direction, while the right, which we followed, took a south-easterly course through a flat country, in which rice, sugar, and maize are cultivated; and a straight reach of 3½ miles brought us to Twa-tu-teen, a large village, where the stream trended to the south, and after another mile and a half we arrived at Mbang-ka.

This is a large town, situated on the river side, and abounding in the narrow and unsavoury streets before described, one side being covered over with a kind of arcade, and the other side open, but by far the dirtier of the two, being chiefly occupied by pigs and children, both of which swarmed everywhere. Accumulations of filth lay about at the very doors of the inhabitants, and it was no unusual sight to see women adorned with bright and gaudy finery sitting within a foot or two of a pool of seething filth enough to breed a pestilence. Chairs or sedans were to be obtained here—rickety vehicles, in one of which I perambulated the town; but in some places the corners of the streets were so narrow that it was with the utmost difficulty that my chair could be coaxed round, and then only by a series of ingenious manœuvres.

Mbang-ka derives considerable importance from the fact that large junks can come up thus far; and one arm of the river, which again divides just above, flows from San-kop-yung, the district which produces large quantities of camphor; and here the junks are loaded with that important and valuable commodity, the source of which is the laurel (Laurus camphora). But the camphor trade is at present of little value to any one, except to those to whom the monopoly is granted by the Chinese Government. The camphor mandarin, as he is termed, who enjoys this monopoly, pays 40,000 dollars per annum into the imperial chest for his privilege, and having obtained the camphor at the rate of about five dollars per picul of 133⅓ lbs., he can then sell it for 27 dollars. One dollar for duties and some other slight expenses increase the cost, and about 10 per cent. of the camphor is lost by evaporation during the transit; for with the proverbial dogged conservatism of their nation, they insist on continuing to pack it in wood instead of stowing it in tin cases, by which contrivance it might all be saved. Still the profits are very considerable, and will probably remain in the hands of the monopolists until some enterprising European merchant shall wrest it from their hands, and open up this important trade to foreign competition.

The branch of the river which diverges above Mbang-ka from that leading to the camphor district is navigable for boats up a series of rapids to the borders of the aborigines' country.

While at Mbang-ka, Capt. Bullock having made an appointment with the chief military mandarin of the district, Ching-yung, to pay him a visit, we repaired to his ya-mun, where he received us with official formality. His residence was situated just outside the town; and our party, including Mr. Gregory, the vice-consul, having reached it, with a procession of chairs at our heels (for we preferred walking, although it was etiquette to go in chairs), we were saluted with three guns as we entered the enclosure. In this enclosure I may here mention that I observed a horse, belonging to the mandarin, of the spotted circus-kind, which seems to be most prized by the Chinese. Mr. Millisch also possessed a horse; but these two were the only horses I saw in all Formosa, though I have been informed that at Tai-wan-foo, the capital, horses are known. Having seated ourselves in the audience-chamber, tea was served in cups of egg-shell china by a number of attendants, as soon as they had succeeded in chasing out the ragged crowd which had curiously followed us into this *sanctum*. The mandarin was decorated with a clear blue button and peacock's feather, and appeared to be an intelligent and rather superior man of about 35 years of age. He conversed freely through the medium of Mr. Gregory, who acted as interpreter; and after having remained some 20 minutes we quitted the place with the same formalities as on entering, the mandarin having first accepted Capt. Bullock's invitation to visit the ship at Tam-suy, next day, which happened to be her Majesty's birthday.

The day following, therefore, the 24th of May, we were prepared to receive his promised visit, and to show him the manner of decorating the ship in honour of that occasion. As usual, at eight a.m., the ship was dressed out with flags, &c.; and Captain Bullock having invited the European residents to dinner, a long table was prepared on the quarter-deck, and a stage erected at one end, upon which the bluejackets were to enact a play which they had got up among themselves. It was not, however, till nearly five o'clock that the mandarin arrived, accompanied by the subordinate mandarin of Tam-suy, his secretary, the consul's linguist, and a crowd of attendants. They seemed much pleased with everything they saw, and minutely examined all the principal arrangements of the ship, particularly the guns; so that time drew on, and the dinner hour (half-past six) was getting very near. The Europeans began to arrive; but our Chinese friends as yet showed no signs of bringing their visit to a termination. Under these circumstances Capt. Bullock, who was quite willing that they should remain, asked them to be his guests for the evening. Chinese politeness, according to the rites, should have declined adding four impromptu guests to an already full table; but undisguised interest and curiosity seized upon the opportunity, and they accepted the invitation without hesitation. Mr. Gregory, the vice-consul, sat with them and acted as interpreter; and as they had their own attendants they fared as well as they could desire. They seemed to appreciate the champagne and other beverages of an English dinner, and did full justice to the viands, even using knives and forks. Dinner ended, and some speeches following, they were politely listened to; and when at length it was announced that the curtain was about to be drawn up, they still kept their seats. The play was "Thérèse," a tragedy of course, for sailors always select something serious and lugubrious, the most affecting parts being, of course, those where it was most difficult to avoid hurting their feelings by a burst of laughter. The Chinese looked on to the end; and even afterwards, when some songs, sailors' hornpipes, &c., followed, they remained politely attentive; and it was only when one of the Chinese servants was forced upon the stage to sing a stave in the real falsetto, singsong, Chinese style, that they allowed their gravity to forsake them, and fairly joined in the laugh which the absurdity of the thing universally raised. Late in the evening they rose to leave, with many expressions of gratification, and three guns saluted them as they went ashore in the captain's gig.

The Tam-suy mandarin, whose name was Lim-ching-fang, wished to have invited us to a return dinner; but unfortunately our plans did not admit of longer stay, and preparations were made the following day for taking the ship round to Ke-lung.

The Sulphur Springs near Tam-suy.

CHAPTER V.

FORMOSA (*continued*)—FROM TAM-SUY TO KE-LUNG.

The Sulphur Springs near Tam-suy; approach to them; their present condition; effects on Animal Life—Preparations for River Voyage—Village of Pah-chie-nah—Arrive at Sik-kow—Bivouac at Chuy-teng-cha—Birds on the route—Rapids—Population—Domestic Animals—Arrive at Liang-kha—Descent to Ke-lung—Character of the People.

HAVING heard of the existence of some sulphur-springs in the vicinity of Tam-suy, I was glad of an opportunity of visiting them; and accompanied by Mr. Lessler, of Tam-suy, who kindly lent me his boat for the purpose, we devoted one of the days of our stay to a journey thither. The locality of the springs is among the hills, about equidistant from Tam-suy and Mbang-ka, and we approached them by taking the left-hand branch of the river, where it divides at Kan-tow. They are highly interesting from a geological point of view, indicating, as they do, the existence of volcanic action near the surface in this part of Formosa—a circumstance which we might have been led to expect from the frequent reports of earthquakes, though none occurred while I was in the island.

These sulphur-springs are not the only springs of the kind in those parts; others are indicated at no great distance. The road to them from the spot where we left our boat ran through a beautiful and highly cultivated district. Besides numerous padi fields situated upon the hill-sides, and ingeniously irrigated by a series of platforms, down which the water flows from one to the other after the manner of the cascades of St. Cloud, a remarkable feature is an immense pineapple-plantation of many acres in extent, so that the verdure of these hills leaves one unprepared for the fact of subterranean heat finding a vent in such close proximity.

On the road we were accompanied by a number of children, who for the reward of a few cash, darted out in forays upon the coleopterous insects of the surrounding country. They brought us splendid longicorns, especially the white-spotted Cerosterna punctator, and the equally handsome Batocera Germani, the first of which we had found in some profusion on the shrubs among the rice-paper hills at Tam-suy. Perhaps the most numerous beetle was a small metallic blue Popilia, and almost equally common was a fine species of green Euchlora, among which was here and there a bronze Mimela of smaller size. Many beautiful yellow Cassidæ were among them; but all partook of a Chinese character and facies.

About halfway up the ascent we crossed a stream having the character of a mountain torrent, the stones at the bottom of which were covered with a deep green deposit, very copious in the quieter and more sheltered spots; and upon dipping one's hand into this stream, the temperature was found to be too high to allow it to remain there. At this point it was about 130°; but higher up it could be seen steaming, notwithstanding the tropical heat of the day.

This stream does not appear to flow directly from the sulphur-springs above, but probably from some subterranean source connected with them. The channel leading down directly from the springs was quite dry, though it bore evidences of having been, comparatively recently, the theatre of similar exhalations. The rocks over the opposite side of this ravine were lofty, and cropped out boldly, striking south-east, and dipping down to the north-east in the direction of the springs. At this spot they had a bleached appearance, visible from a distance, precisely similar to that exhibited at the active springs. They bore, however, at that moment, no other sign of their past activity; but, on a near approach to them, a very perceptible odour of sulphuretted hydrogen was smelt, and the rocks themselves appeared to have had their surface disintegrated by the action of the steam.

A short distance above this spot we reached a *cul-de-sac* in the hills, bounded on the right by bold bare rocks, having the lithological characters of a coarse calcareous grit, and dipping about 15° to the north-east. This was the spot occupied by the present active sulphur-springs, and was of small extent, embracing not more than two acres of ground, whose desolation formed a very striking contrast to the verdure on nearly three sides of it. This spot was perfectly barren, and was filled up with low hillocks of friable rocks, loose stones and débris, having the character of a moraine, and interspersed at irregular intervals with shallow pits or depressions, containing mud and sand, and sometimes foul, muddy water. From cracks and fissures in these depressions arose clouds of steam, and yellow patches of sulphur were visible from a distance.

At the time of my visit, in the middle of June 1866, there were seven or eight springs in a more or less active condition, from which clouds of superheated steam arose, either by a small round hole, or narrow fissure, or by several such apertures. The rushing steam produced a loud noise, like that accompanying the blowing off of steam from a boiler; and above the fissures was a quantity of sublimated sulphur, adhering to the rock in acicular crystals, forming, about the most active spring, a bright yellow patch which was visible from a considerable distance. It was no easy matter to reach the sublimed sulphur, for, on a close approach to the spot, a jet of hot steam made it necessary to withdraw, and warned us that a nearer approach was dangerous. I managed however, with the aid of a stick, to procure some from the crevices in and around which it was deposited. Most of the springs were dry; but one rose through muddy water, which bubbled up in a series of rapid explosions, carrying the boiling water, sand, and mud five or six feet high, and splashing it all around.

It is evident that the degree of activity of these springs is very variable, and that at the time of my visit they were in a comparatively quiescent state. The jets of steam were isolated, and a comparatively small portion of the two acres, at which I estimated the area of grey barrenness, was in an active condition. Numerous pits which had evidently at some period sent forth their jets of steam were perfectly quiet, and stones coated with sulphur scattered among them showed their occasional activity. Moreover, the edge of the level, where it began to descend down the ravine before mentioned, was covered with a thick crust, which had evidently been at one time in a semifluid state, and had slowly flowed, a viscous mass, over the edge, and now had the appearance of dried asphalt. This was doubtless the remains of mud, through which the sulphur rose, such as we still saw in some comparatively small pools, but which at one time had been in sufficient quantity to rise above the general depression and run over the edge into the ravine.

The sulphur appeared in all cases to be deposited in a perfectly pure sublimed form; nor was there any smell to be detected in the active springs themselves. The steam is laden with the element in a dissolved condition, and deposits it in pure crystals upon any substance with which it comes in contact. The effects produced upon the exposed rocks were in all cases due to the disintegrating and bleaching effects of steam; and the smell of sulphuretted hydrogen was most perceptible in a spot where the rocks had been disintegrated, but where there was no sign of present activity.

It has been supposed that the locality is very fatal to animal life, from the presence of sulphurous vapours,—that it is a sort of Avernus, destroying birds and insects which pass in its neighbourhood. But I cannot endorse this view. I myself observed birds and insects flying over it with ease and impunity, nor was any noxious smell elsewhere perceptible. Any ill effects could be produced only by the direct action of the *steam*, with which the *sulphur* could have little or nothing to do; and if any corroboration of this were required, it need only be mentioned that the patch occupied by the sulphur-springs is immediately surrounded by the brightest verdure, and a stream of clear water runs along its edge, and alone separates it from padi fields in the most green and healthy condition.

At the present time no attempt is made to obtain sulphur from this prolific source. Although it can be obtained at the rate of 45 cents per picul of 133 lbs. (about 2s. per cwt.), the Chinese Government stupidly and obstinately forbid its being worked. Still, sulphur has been largely obtained from these springs under the rose, or by means of a bribe, and it yet remains for European enterprise to open up so important and probably almost inexhaustible a source of this valuable material.

On the 25th May, Captain Bullock having decided to take the ship round to Ke-lung harbour, on the east coast, I made arrangements to proceed overland and meet her there. The journey across the country could, as I learned, be performed almost entirely by boat, with no other difficulty than some rapids in the higher part of the river. It was, moreover, short, and was not unfrequently performed by two or three gentlemen who carried on the occupation of merchants either at Tam-suy or Ke-lung, and had often occasion to communicate personally with one or the other town. Having therefore obtained the necessary information from them, Mr. Sutton, the chief engineer of the "Serpent," and myself, proceeded at once to make our preparations for the voyage.

Having obtained a *sampan*, or native boat, with three men, we placed in it provisions for two days, camera, collecting apparatus, &c., intending to proceed leisurely. The boat was a flat-bottomed one, adapted for the peculiar navigation, about 20 feet long and six feet wide, covered with a bamboo awning, and having a grass mat at the bottom; and, with the aid of a large mat-sail and a sea-breeze, we rapidly proceeded up the Tam-suy River, soon arriving at the spot where it first divides at Kan-tow. From here we followed the right-hand branch which flows east by south through cultivated fields, in which we occasionally met with patches of Boehmeria nivea, and small groves of betel-palm (Areca catechu); but the characteristic tree of the banks here, as everywhere along the river, was the bamboo, whose graceful and feathery foliage gave a great charm to the scene. On the north-east side were numerous hills, of heights varying between 1000 and 1500 feet, amongst which are situated the sulphur-springs, already described. A little more than three miles brought us to the village of Pah-chie-nah, which is more airy and cleanly than either Mbangka or Hoo-wei, and possesses an excellent market-place, though the inhabitants appear to be of the same poor class. Numerous duck-boats were met with on these banks, which bring some couple of hundred ducks to a feeding-ground, where they are turned loose to spend the day under the charge of a lad, who acts as duck-herd. They keep close together all day, so that they might all be covered with a blanket, and at night are conveyed in the boat back to their pens. Another feature of the route was the Chinese water-wheels for irrigating the fields, in which three or four Chinese are constantly at work, treadmill-fashion.

At sunset we moored our boat a mile beyond Pah-chie-nah, in a bend of the river and at the foot of a hill which commanded a magnificent view of the noble range of mountains running from north to south of the island, and which the setting sun lighted up gloriously. On the opposite side of the river, upon a steep rocky bank, was a house, outside of which sat a family of Chinese of a better class, the head of which having, somewhat to our surprise, leisurely examined us with a good double field-glass, made signs for us to go over and *chin-chin* with them. We accordingly did so, and, having partaken of their tea, offered them some of our own provisions, with which they appeared much interested, particularly the white bread, though the loaf-sugar seemed most generally appreciated.

We slept in the boat, the night being brilliantly fine, a strong dew falling towards sunrise, and the stillness being broken by the croaking of frogs, the chirping of cicadas, the occasional leaping of a large fish in the stream, the passage of boats up the river, and the distant creaking of a waterwheel which appeared to be in action all night long. A strong tide was flowing; but the water appeared perfectly fresh to the taste, even at the flood. We had agreed to keep watch and watch during the night, and I most religiously kept awake during the first hours, listening to these various sounds. When, however, my turn had passed, and after a short nap I awoke, I was not a little disgusted to find my companion snoring instead of watching. But there did not appear to be any real reason for the precaution.

The following morning, after taking some photographic views, capturing some of the beautiful butterflies and beetles which, especially the former, abounded on the hills, we proceeded on our journey. The thermometer being at 89° in the shade, we were glad of our bamboo awning; and there being no wind and a strong ebb tide, we made but little progress for some time, moving slowly by a very meandering course through a highly picturesque country. Hills of varying height rose on either side, usually covered with vegetation, and occasionally opening and showing green padi fields; while in front an abrupt and very remarkable long stratified hill occupied a conspicuous part of the landscape, and this we gradually approached till we reached the town of Sik-kow, behind which it was situated.

Sik-kow is similar in character to the other towns on the route; but the streets are wider than those of Mbangka or Hoo-wei. The inhabitants, however, did not give us any notion of their being more simple or primitive on account of their comparative seclusion, but rather the reverse. A noisy crowd followed us through the streets, some members of which appeared to incline to impudence, and one man seemed by his loud talk and gestures to be attempting to incite others against us, while the general greeting of "*hwan-ha*" (foreigners) was heard no less here than everywhere else on the route.

Leaving Sik-kow, we proceeded eastward through similar scenery, increasing, however, in its striking character, for some six miles further. A little beyond Sik-kow on the left bank, a bed of large oyster-shells, some of them eight or nine inches in length, and having a close resemblance to, if not identical with, the recent Ostrea canadensis, arrested our attention. They were imbedded in stiff blue clay in the river's bank, and immediately overlies a thin seam of an inferior coal, which cropped out beneath. The bank (which, as in most other places, was perforated with the innumerable holes of freshwater crabs), including clay, shells, and coal, was about four feet high above the water's edge, and the bed extended about 100 yards in length.

We arrived at the town of Chuy-teng-cha at nightfall; and here, as its name implies, the tide-way ends. As it was dark we did not land, but proceeded a little further, and passed the night in a small bay at the foot of the rapids. Numerous boats upon the beach and many in motion seemed to show that this was a busy town of some importance; and by questions put and answered, as we passed, in which we could hear from time to time the word "hwan-ha," we knew that the people were discussing our movements and the kind of freight our boatmen had under their charge. We had no fear of them, however, for they turned out to be excellent fellows, good-tempered, willing, and obliging, and mightily amused at all our proceedings—one of them, in particular, laughing from morning till night.

On the second night, as before, we were tormented by mosquitoes, which made it difficult to obtain any rest; while the close heat of the atmosphere made us wish to divest ourselves of some of our clothing, a proceeding forbidden by the tormenting insects. Frogs and cicadas again kept up a serenade all night; and a nocturnal bird sang a harsh song in some trees upon a cliff opposite. I could not get a sight of this bird, whose four notes somewhat resembled the creaking of a wheel; the last two notes being often repeated twice. As soon as dawn began to appear it flew away, and I heard it no more. At the same time two or three large bats, which at first in the twilight I mistook for owls, flew home to their retreats with a loud croak.

As soon as the sun arose, pheasants began to crow upon the fern-covered hills, and we heard and saw several during the day; but, although we landed for the purpose, we were unable to get a shot. But by far the commonest bird we met with throughout was a black bird—whose feathers, however, had a rich green gloss—about the size of an English ousel, with a long forked tail and whitish rump. This bird made a harsh note not unlike a jay. This was the Black Drongo, Dicrurus macrocercus of Latham; they were visible everywhere along the banks, usually in pairs, seldom flying over the river, and often perched upon the topmost spray of a bamboo in a conspicuous position. I procured the nest and eggs of this bird. The nest was made of dried grass and cotton-grass, simple in form, and situated upon the bough of a tree about 15 feet from the ground; the eggs were three in number—pinkish, with sparse umber spots and blotches, particularly about the larger end. The other birds I noticed were doves of a small species, kingfishers, pied wagtails, grey shrikes, and a small short-tailed bird (Cotyle sinensis), with the habits and character of a sand martin. Early in the morning, a lark (probably Alauda cœlivox) singing in the fields could scarcely be distinguished from the English skylark, and another bird's song reminded me greatly of the English song-thrush. A second thrush-like bird also was singing, as well as the sprightly little Prinia sonitans; but not more than half-a-dozen birds could be said to be in song here, at a time when nearly thirty would be enlivening the woods and groves of England.

Having passed the end of the tide-way, the remainder of the journey was made through a series of strong rapids, up which it was necessary to drag the boat by main force. They commenced immediately from our resting-place of the previous night, and our boatmen jumped out of the bows, and passing a bamboo across them pushed one on each side, while the third pushed behind, and thus our flat-bottomed craft moved up the incline into a reach of deep water. This proceeding was repeated perhaps a score of times, the intervening reaches being bounded by very beautifully wooded hills, with precipitous rocks dipping to the water's edge about 15° to the east. Many beautiful secluded retreats were thus passed, generally, however, with signs of life near them; for it is remarkable how densely populated this side of the island appears to be—nowhere could we go without meeting Chinese in some form or other: in the quietest and most retired spots a cottage might often be descried upon close inspection. If we wished to shoot a bird among the brushwood, we were most likely to find a group of women and children peering at us from behind; if it were on the bank, some fishermen at work, or lads wading in the mud for shell-fish, or women washing in the stream, were sure to be there, so that it was never safe to shoot, except at the upper part of the trees. Ferries were numerous, and generally at work as we passed; water-wheels were met with at every turn, generally worked by three men, or two sets of three; children leading water-buffaloes on the bank were frequently seen, and the unwieldy heads of these animals often peered at us above the water with a mingled expression of curiosity and stupidity; and even in the midst of the stream were Chinamen and boys, sometimes stark naked, but more frequently with something about the loins, dredging for shell-fish and crabs in the river. The shell most commonly obtained in these situations was a dark costate species of Cyrena; but in the markets two other species were equally abundant as articles of food, viz. Cytheræa petechiana, and a species of Tapes. A long black Modiola (M. teres) was also largely eaten. But everything is fish that comes to the Chinaman's net, and he is always at work, even in the most unpromising situations, to earn a livelihood in a mud-bank, or a sand-flat, or up to his neck in water in a river. Population teemed everywhere, and, while in England we might have walked for miles without meeting an individual, we were scarcely ever out of sight of some human being in this part of Formosa.

The houses were built of mud and thatched, occasionally

more substantially of brick and tiles, but usually of grass and reeds, arranged in tiers, and plastered over with mud and cement,—the floor, even of the better houses, of mud or earth,—the roofs, often crescentically gabled, giving the towns a very characteristic appearance. In the poorer houses in villages, the pigs and fowls made themselves quite at home in the interior, and I have seen a large cesspool only partially separated from the dwelling-room. Pigs, fowls, ducks, geese, and buffaloes, were the only domestic animals, if we except the dogs and cats. The cats were mostly of the Malay breed, with a short broken or twisted tail, and usually tortoise-shell in colour; the dogs most commonly black, seldom white, of an ugly mongrel appearance, about the size of a pointer; they barked vigorously as soon as they caught sight of the foreigner, though there was no fear of their biting, provided we possessed a stick, for they were most arrant cowards. Horses and asses were unknown, and humped cattle, of a small size, rare.

At length we entered a narrow gorge of rocks, which only left room for two boats to pass one another, and warned us that the aquatic part of our excursion was at an end, and in a few minutes we were in the midst of a number of boats the counterparts of our own, which completely lined a beach about 100 yards long, scarcely leaving space for the painted nose of our own craft to insinuate itself between them. Here were clustered some houses forming the village of Liang-kha, about three miles from Ke-lung, where the river we had ascended abruptly terminated on the shoulder of a hill, up which we had risen by a series of rapids, another and a smaller stream branching off from the same spot, and descending the other side towards Ke-lung.

Having placed our gear in a chair obtained from Ke-lung,

we proceeded on foot through a pass on the hills, meeting on the way numerous coolies transporting goods of various kinds from Ke-lung. Some carried heavy bundles of dressed hemp; others, barrels of dried flying-fish of a large size. A sudden turn of the road brought us in view of a splendid panorama—the valley, town, and spacious harbour of Ke-lung, forming altogether a fine picture. On the densely wooded knolls in the valley, tree-ferns were conspicuous; the sandstone hills on the left dipped in long stratified lines to the south-west; and outside the harbour, in which three square-rigged ships, as well as numerous junks, were lying at anchor, stood like a sentinel an abrupt rock, 600 feet high, known as Ke-lung Island, and bearing some resemblance to St. Michael's Mount. On the right was the interesting coal-region, which renders Ke-lung so important a port, in which good anchorage and plenty of fuel may be always readily obtained.

Descending into this valley we passed through the town of Ke-lung, paying our sole visit to it on this occasion. It seemed larger and more open than those we had previously seen, but was inhabited by the same class, who indeed gave us an unfavourable impression, by detaining us at the landing-place until we satisfied their exorbitant demands, urged with an unpleasant degree of noise and tumult. We afterwards learned that they had on more than one occasion threatened the life of one or two Europeans who were resident here, and who by maintaining an independent demeanour, and refusing to succumb to their prejudices, had rendered themselves obnoxious. Like the Chinese generally, however, they require to be dealt with firmly; and the only way to establish oneself in security among them is to show them a bold and determined front—taking at the same time proper means to avoid public collision or private revenge. The vice-consul, however, a kind and well-intentioned man, possesses but little of this determination, and is unfortunately too ready to humour their prejudices, and show respect to their most outrageous feelings and wishes, which have more than once bred riot both at Tamsuy and at Ke-lung.

At length we got clear of the turbulent crowd, and having placed our *matériel* in a boat, we found the "Serpent" at anchor near the mouth of the harbour, and at a considerable distance from the town, which was the principal reason why we never returned to it, although it is to be regretted that no photograph was taken from the hills behind, which commanded so unusually fine a prospect.

CHAPTER VI.

FORMOSA (continued)—KE-LUNG.

Prevalence of Sandstone—Formation of the Harbour—Caverns—Village Populations—Modes of Fishing—Sandstone Peaks and Images—Rising of the Coast—The Coal Mines; mode of Working; value of the Coal; geological position of the Beds; burning Properties—Petroleum—Marine Animals of the Shore—Peronia—Aplysia—Nudibranchs—Creseis—Singular shoal of Stephanomias.

FINDING good anchorage in the harbour of Ke-lung we remained there for some weeks, or rather we made this place a starting point for some interesting excursions, returning to it again during the interval, and allowing thus some opportunity for geological and natural history investigations.

On this side of the island sandstone prevails, and the whole environs of the town of Ke-lung are of that rock, which extends from Masou peninsula, north of Ke-lung, to Petou promontory on the south and east. The section of the coast between these points exhibits inclined beds of red sandstone with an average dip of 16° or 17° to the southeast, the weatherworn outcrops producing an undulating country. The hills at the back of the town of Ke-lung are also of the same formation, and have a similar dip and strike. The harbour of Ke-lung is a spacious excavation in these sandstone strata, the navigable entrance being narrowed by a low flat sandstone table ten feet above high-water, called Bush Island, on the south side; between which and the mainland is also a larger island, which has evidently been separated from it by the bursting of a narrow passage, and its subsequent gradual wearing away by the sea, which washes through at high-water. This is Palm Island, upon which, however, no palms grow; but a few Cycads, which have probably been mistaken for them.

Ke-lung harbour presents many remarkable and interesting features. The north side is picturesquely indented, and more or less covered with luxuriant foliage; but the south side, where the ascending strata are abruptly broken off, presents a beautiful succession of rounded knolls, separated by narrow valleys and steep-sided ravines; the whole being densely clothed with trees and verdant underwood, in which occurred yellow Cassidæ of various species, and great numbers of a beautiful fringed land shell (Helix trichotropis). Several caverns exist upon this side of the harbour. The largest and most remarkable of these was reported to be of very considerable extent, and mysterious tales were told of the difficulty or impossibility of exploring it; we therefore determined to make the attempt. The entrance was prettily ornamented and overgrown with ferns, lycopods and begonias; it faced the mouth of the harbour, and was lofty and spacious, having a sandstone roof above of 50 or 60 feet in thickness. The main cavern was arched and symmetrical; but we soon found that this part at least was very limited in depth, for having penetrated about 50 yards we arrived at its extremity. It was rather damp, and the floor was of hard sandstone, presenting no indications of any deposit in which one might look for organic remains with any chance of success. In the left wall, however, we found a narrow fissure, which was the really unknown portion, and being provided with a magnesium lamp, we squeezed ourselves into this cleft, and crept along it with difficulty for 70 or 80 yards. One stoutish gentleman of our party fairly stuck in the middle, and was only hauled out with difficulty by the combined efforts of a blue-jacket before and behind. This fissure was very damp, and at length terminated in a small irregular chamber, beyond which we could see no passage. Its proportions were well seen by the aid of the magnesium light, which so illuminated the vault, that the gentleman before mentioned, who was wedged in midway, flattered himself that if he could once get out of his difficulty he should find himself in broad daylight.

While examining the walls by the aid of this light, I found they were tenanted by some spiders and crickets, of the latter of which I with some difficulty secured specimens. They proved on examination to possess perfect eyes, although the place is of course totally dark; and it is therefore to be concluded that they had simply crept in from the exterior, though what could be the inducement, or upon what they subsisted, it is difficult to imagine—for the rocky walls were perfectly bare, and the whole intervening distance was dripping with water. Returning to the main cavern, I was curious to see if they also existed there; and at the extremity of this I also discovered the crickets upon the damp wall.

On either side of the harbour are several villages, inhabited by a poor fishing population. These villages give a lively aspect to the spot. One of them, close to the mouth of the cavern, being near the landing-place, we often visited, and soon became acquainted with the entire population, whom we employed in collecting shells, &c. Had the district been a rich one in these commodities, we should doubtless soon have made a good collection, for the younger part of the community, both boys and girls, soon found that they could raise a little money in this way; and every time we appeared on shore we were speedily surrounded by an eager crowd of half-naked and tolerably dirty urchins, who pressed upon us the common but pretty cowries, Cypræa pellis-serpentis and C. annulus, &c., and all manner of trash. The boys were forward enough, but the girls were very timid, and for a long time would only hold out their hands at a distance to show they were anxious to trade, but afraid to come too near the *Hwan-ha* (foreigners). From the miscellaneous collection thus presented to us, we selected a small number, for which we had a fixed price of a few *cash*—a most convenient medium of exchange—for inasmuch as 1000 go to a dollar, two cash for a cowry, a helix, or a beetle, while it added considerably to their exchequer, at the same time did not threaten to ruin ours. The inhabitants of these fishing villages were Chinese, and therefore not idle. The girls and younger boys were daily out, as long as the tide permitted, among the rocks gathering shell-fish, and it was not uncommon to see them up to their necks in water, collecting what forms to them a very important article of diet. The men were employed in the fishing-boats, or in hauling the seine, which took place chiefly on the sandy beach of the south side; but although the seine was very extensive, and the operation of pulling it a very laborious one, the result did not seem to be at all adequate. I went ashore on one or two occasions during the haul, in search of fish, but did not succeed in getting anything but very small sprat-like fishes. Nothing larger appeared to be caught. Every evening, too, at dusk, numerous boats pulled out from the town to the wider part of the harbour, and after dark a number of blazing torches spread a lurid light over the water. Curious to see what they were doing, I one night took a boat and rowed amongst them. In each boat stood a man at the bow, holding a bundle of small bamboos, which blazed so brightly that I was hardly convinced that they were not dipped in oil, until an examination proved them to be dry. This flaming torch, with which about three boats out of four were provided, served to attract the fish, and when the boat was thus surrounded with fish, a signal was made to another boat unprovided with a light, which coming up, drew a net around the illuminated boat, and thus secured the shoal. This was done with great shouts and noise, which we often heard, lasting far into the night. I boarded one of these boats, and saw them haul the net; but the produce seemed to be entirely confined to a small white fish like whitebait. The fishermen were civil enough, although we rowed right in among their operations; but they did not appear at all anxious to dispose of the fish, which were sweet-tasted, and if cooked at Blackwall would probably rival the real whitebait.

The effects of aqueous action upon the sandstone rocks are very conspicuous in some parts of Ke-lung harbour. Near the cave before-mentioned, and immediately upon the verge of high water, is a tall isolated sandstone rock, having precisely the appearance of an old ruined castle, and appropriately named Ruin Rock, which forms an excellent landmark by which to anchor a ship. The harder layers of sandstone having defied the effects of weather and the spray which is dashed up during the north-east monsoon to which the harbour is exposed, the softer portions have at the same time been more or less excavated, leaving a mimic resemblance of the ruined chambers of a three-storied building. But the most curious and extensive effects of the direct action of the sea are to be found at the entrance of the harbour on either side. That on the north side is called Image Point on the chart, but the south side is even more remarkable, and no less deserves this name, while the effects are upon a larger scale.* Crossing over the narrow sandstone platform connecting Palm Island with the mainland, and which is covered at high water, I found myself in an extraordinary spot, where the soft sandstone has been worn away by the force of the waves into a variety of fantastic forms, for the most part resembling gigantic mushrooms—huge stalks, 10 or 12 feet high, bearing vast balls of harder material upon their summits, like immense ninepins; hills with excavated flanks, and harder knobs and ridges, over the foremost of which the waves were dashing, sending up the spray 50 or 60 feet high, although the sea was comparatively calm. Some of the heads of these huge mushrooms had fallen off, and remained as great round blocks with hard ridges, such as are often seen, but whose history could here be distinctly traced, as could also a further step in the disintegration of the beach;—for in many places round, deep holes were bored in the solid rock, which were evidently produced by one of these hard heads resting upon a softer spot, where it had been twisted and whirled about by the waves, wearing and boring its bed as though with an auger, sinking deeper and deeper, until at length it was itself worn away and dissipated by the long-continued grinding action, leaving a clean-cut deep hole in the rock from a foot to a yard in diameter, but containing nothing but clear sea water.

I have little doubt that the harbour of Ke-lung is slowly rising, though I have not sufficient data to show the rate of elevation. The evidences of this elevation are to be

* See Frontispiece.

found on both sides of the harbour. Blocks of worn and washed coral strew the beach on the north side, and lie about confusedly at high-water mark in the neighbourhood of Ruin Rock. Similar washed coral blocks lie on the beach between tide-marks on the south side, viz. on Palm Island. The sandstone platform between Palm Island and the mainland, which presents every appearance of having been excavated by the sea slowly forcing a passage through, is now very little below high-water mark; and above the sea level the sandstone rock bears plain indications of having been washed and worn by the waves where vegetation is now growing. Beyond the present limits of the harbour, the level plain at the back of the town shows that the sea once extended farther among the hills; and the inner third of the present harbour is so shallow as to be a mere mud flat at low water. Quite recently the middle third has become too shallow for the anchorage of large ships, such as had previously found sufficient depth; but this fact may be due to the evil practice of throwing ballast into the harbour to save the trouble of carrying it ashore; for, although the Chinese are industrious enough to work when necessary, they have but little conscience; and if engaged to unlade a ship in ballast, they will do so, but will drop it overboard at the nearest convenient spot, as I have seen them do, without the slightest consideration for the deterioration of an anchorage, or the shoaling of a sheltered landing-place.

This part of Formosa derives commercial importance from the existence of coal-mines, which are possessed and worked by the Chinese authorities. I visited these mines, which are situated about a mile and a half to the eastward of the town of Ke-lung, on the sides of the hills bordering on Quarse-kau Bay. Being in communication with the owners of coal depots for the purchase of coal for the ship, we were brought into contact with a civil Chinaman, who was acting as compradore for Messrs. Lessler and Hagen of Tam-suy. This man spoke and wrote excellent English, having been educated at the English school in Penang, and when subsequently he superintended the delivery of the coals on board ship, the sailors were not a little astonished, and stood around open-mouthed, to see a smart young Chinaman with pig-tail, long silk coat, thick-soled shoes, and about whose nationality there could be no mistake, sitting at a table on deck and writing an elegant, free, commercial hand, while he communicated with the officers in fluent and grammatical English. This man politely lent us his gig and two rowers to conduct us to the mines. The two men were very good-humoured, particularly the younger one, who laughed immensely at everything we said and did. Having rowed us nearly up to the town, the harbour getting very shallow as we proceeded, so that at length only a narrow channel between two mud-flats approaches the town, we entered a small, muddy creek, with so little water that our boatmen had several times to jump out and pull the boat along. The hills were beautifully wooded, and the glen narrowed as we proceeded. At length, quitting the boat, we ascended a slight elevation, passing a range of red sandstone hills, which formed a series continuous with those seen at the back of the harbour, and which dip on an average 16° or 17° to the south-east. The weather-worn outcrops of these strata produced the undulating country in which I now found myself, and in the depressions of which the coal appears to have been deposited. We now entered a *cul-de-sac* in the hills, and, descending from the path into a ditch, I stood at the entrance of the workings, which consisted of two small caverns at right angles to one another, hewn directly into the coal seam, which was 2½ feet in thickness at its outcrop. The seam rested upon a thin bed of stiff, whitish clay, and was covered by a bank 40 or 50 feet high composed of rubbly clay with stones, on the face of which small bushes were growing. Out of these caverns a dirty stream of water was flowing, ankle deep. The working was nearly level, and the roof so low that one could only get along by bending nearly double. There was nothing remarkable in the interior; the workmen, all Chinese, were in a state of perfect nudity, and after a painful and very dirty walk of about a quarter of a mile, we emerged at another part of the hill.

These mines appear, therefore, to be worked in a very primitive manner. No shafts are sunk, nor is any machinery employed, but the coolies pick the coal and convey it out of the working in small baskets, and in almost infinitesimal quantities at a time. It is placed in boats and conveyed to the harbour, where it is deposited in the coal-stores situated upon the southern side—mere accumulations of coal purchased by English and other merchants, and from which ships are mostly supplied. These stores have no covering, nor any protection whatever from the weather, and the coal therefore is apt to deteriorate if kept there long. The mines themselves are exclusively worked under the Chinese authorities, and by Chinese coolies, foreign interference or possession being jealously guarded against: the consequence is, that their resources are both undeveloped and unknown. It is impossible to judge of their extent beneath the soil, because no shafts have been sunk, and no tentative efforts in the shape of borings appear to have been made. The wonder is rather that so much is produced by the industry of the coolies; its comparative cheapness is owing to the low value of coolie labour, added to the absence of expensive outlay in the working of the mines.

The coal resources of Ke-lung have only recently been made known. In 1857 it was stated that "owing to the prohibition by the authorities of Formosa against the export of rice, vessels arrived at Amoy loaded almost entirely with coal, at about $1\frac{1}{4}$ dollar (5s. 6d.) a ton;" and it was further said at that time that arrangements might be made for the formation of a stock for the supply of Her Majesty's vessels on very favourable terms.

In 1858 H.M.S. "Inflexible" received coals at Ke-lung at the rate of four dollars (17s. 6d.) per ton. H.M.S. "Serpent," during the year 1866, was coaled at the rate of 16 dollars the hundred piculs, which is somewhat less than three dollars (13s.) the ton; and for this price we selected our coal from the depots, and it was brought alongside and deposited in the bunkers. When we finally left Ke-lung, there were seven ships in the harbour—Hamburg, Bremen, Prussian, and English, receiving coal either as cargo or for consumption.

The position of the coal-bed of Ke-lung proves that it is of comparatively recent formation. It lies apparently quite superficial; and, although it would undoubtedly require a closer and longer study than I was able to devote to it, in order to prove its exact geological relations, especially in the absence of any subterranean workings in the form of shafts or borings, the position of the worked seams is undoubtedly superficial to the sandstone. How far down the coal seams are believed to penetrate I was unable to learn, for they are in the hands of Chinese proprietors, and all the workers are Chinese, with whom I was unable to communicate directly;

while the few European merchants who are interested in the produce were not scientifically acquainted with the district.

With regard to the quality of this coal, it has properties which favour the supposition that it is a recently-formed deposit. The first account of it made public was issued from H.M.S. "Inflexible," and the chief engineer of that ship published an account of his experiments and steaming results with it, in the *Nautical Magazine* for 1859. This account, however, is strangely at variance with our experience of the coal. In general terms the verdict given by him was that it was "good for domestic purposes and for steamers making short passages; but it consumes rapidly, and makes much smoke." Although, however, this general statement nearly coincides with what we found to be the case, it is not supported by the elaborated and tabulated results published in the *Nautical Magazine*.

The Ke-lung coal is of very light weight; it burns very rapidly, and it gives out a very great heat—so much so, that it readily sets the funnel on fire. It is extremely dirty, and the combustion is so imperfect, that a vast number of blacks of a soft and soiling character are produced, and fall all over the ship. The flues also rapidly get very foul, requiring frequent attention and cleansing. It leaves no less than 50 per cent. of ash, so that although it appears cheap, it is not really more so than other and better coal, which has more substance and less waste. For it is evident, that if Ke-lung coal were but one-half the price of Welsh, and that Welsh did twice as much work, the latter would be cheaper fuel; for not only would there be equal horse-power for an equal price, but the superior bulk of the inferior and apparently cheaper coal would entail great additional labour upon the firemen in removing it from the bunkers and feeding the furnaces, to say nothing of the waste of stowage.*

But the worst feature of the Ke-lung coal is that it forms a large quantity of slag, or *clinker*, which sticks firmly to the furnace bars, and becomes so heated as to fuse them. Many of the fire-bars in the "Serpent" were fused in this manner before the load was exhausted.

I was informed that at no great distance from the coal mines of Ke-lung there are sources of petroleum, which are known to some European merchants residing there, who were in treaty for the ground. The Chinese, however, are very jealous in guarding any land which is supposed to possess mineral riches, having an idea that gold is to be found there. So anxious are the present Chinese occupiers upon this point, that in any title of purchase of land there is an express stipulation, that should gold be discovered upon that land, the precious metal should not be considered as included in the purchase, but shall revert to the original possessor of the soil.

The rocks around Ke-lung harbour did not yield a very great variety of animals, although there were some of considerable interest. The sandy beach in some places was entirely formed of minute shells of a great number of species, usually more or less rubbed, but containing a considerable number of tolerably perfect specimens. In the crevices of the coral blocks which strewed the shores, shoals of small and beautiful coral-fish abounded, some of the richest azure blue (Pomacentrus), others striped and banded (Glyphito-

---

* There are some interesting points of resemblance between the coal field of Ke-lung and that of Labuan, on the coast of Borneo, of which an account will be found in Chapter X.

don and Therapon), others yellow, green, red, and various bright colours, and of forms equally various; but unfortunately neither spirit nor glycerine succeeded in preserving their tints. The rocks, where washed by spray, were blackened by the swarms of Ligiæ running nimbly about, exhibiting a bluish metallic tint, which glanced upon their backs in the sunlight. Beautiful purple Echini occupied the hollow places in the sandstone; and great black Holothuriæ, of the kind used for Trepang, lay scattered about in many places, and these, when touched, threw out a quantity of white tenacious threads, which adhered like glue to the hand. The slug-like Peronia was not uncommon, usually found crawling upon the rocks at high water, being an animal that is satisfied with an occasional moistening of the surface. When I kept these animals alive they proved very erratic, and would never remain in the vessel, but immediately crawled out; and I found them from time to time in all parts of my cabin, even some days after I had lost them. I was therefore somewhat surprised, on a subsequent occasion, to find Peronias on the coast of Borneo, on the under side of stones which were immersed in the water. The Peronia is greenish-brown in colour, without dorsal branchiæ, or mantle-tentacles, as in the Nudibranchs, but have two snail-like retractile tentacles on the head, with eyes at their points, and the whole mantle is covered with papillæ, having something of the form of fleurs-de-lis. After the gale which detained us in the harbour, the low cay, called Bush Island, was covered with a fleet of little Velellæ and Physaliæ, which had been stranded by the wind. This island, too, produced a number of beautiful Anemones, botrylliform Tunicata, &c.; but the most remarkable animals there met with were certain Tectibranchs, as they are termed, in which the shell is more or less undeveloped and concealed in the mantle, the gills forming leaflets also under its protection. These were the sea-hares (Aplysia), of which at least two species lived here—one, the most common, of a uniform brown colour; the other, of larger size, marked with sparse black blotches. These animals are remarkable for their power, like the cuttle, of pouring out an abundant secretion of a purplish colour from the edge of the mantle, with which, when alarmed, they stain the surrounding water. Another somewhat similar animal found here was the rich black Coriocella nigra, its flowing velvet mantle entirely concealing its shell at pleasure.

In Ke-lung harbour, although I sought diligently, my pains were rewarded by only two or three species of Nudibranchiata. Of these one was a small blue Doris, on Bush Island; the other two were, however, both new species, and interesting from their extreme beauty. One of these was a Doris of a cream-colour, edged with orange, and covered over the back with rich vermilion marbling. But the last was probably the type of a new genus, its mantle capacious, of a rich variegated rose colour, edged with white, and studded with translucent white spots—the whole body so delicate as to be semi-transparent. Its movements were wonderfully graceful; spreading the broad and transparent mantle out wide on either side, and throwing back its long tentacles, like ears, it swam about with a moderately rapid vermicular but vertical motion, the head and tail being thrown forward till they met above, and then partially thrown back, accompanied by a waving of the mantle from end to end.

Lying for some time at anchor in this harbour, some very interesting marine animals came under notice from time to time. At one time the towing-net would bring up transparent animals which bore a close resemblance to the Cymbulia ovularis, of Rang, whose broad expansive wings, by which locomotion was effected, were placed in a tuberculated and purse-shaped crystal calyx, from which it was easily separable, the whole animal being in some lights invisible but for an oblong black spot in the centre. Another of these transparent Pteropods was the Pneumodermia; but the most abundant and striking was the pretty and delicate little Creseis, with an elegant glassy shell, like an inverted church spire, pointed like a needle at one end, while, from the other, a pair of little delicate wings would keep the calm surface of the water in a constant ripple by their soft flapping to and fro. So abundantly did these little creatures swarm upon some days, that they came up in solid masses, and the towing-net was filled with them in every mesh; so that it was a long task to clear it of the fragile shells.

Beautiful Acalephs, or sea-jellies, too, were among the harbour's inhabitants; ciliogrades, like elegant pink glass flowers, in constant motion, with prismatic bands of cilia playing along the raised ridges of their body from end to end. But even these were hardly so striking as the wonderful influx of Hydrozoa, of the singular genus Stephanomia, that occurred one evening. This happened upon the 18th of June. Although calm, it had been a wet day; yet, in spite of this, myriads and myriads of Creseis swarmed in the harbour. During the day a breeze sprang up, and at times rollers came in; but as the afternoon advanced, the sea became alive with marine animals, including some of the forms I have already described, but chiefly beautiful organisms which most closely resembled the Stephanomia triangularis of Quoy and Gaimard. They were wonderfully sculptured and carved masses of solid jelly, either perfectly transparent, or tinged with pink. They would bear being taken up carefully in a hand-net, and placed in a basin of sea water, but when there, they became absolutely invisible from their delicacy and transparency. When touched they would break asunder into transparent, gelatinous, star-like bodies; so that I was in despair at getting even a sketch of their complicated forms, for they soon melted away into shapeless masses. I endeavoured to preserve some in various substances, but without success, for they immediately fell to pieces and dissolved.

These bodies were solid to the touch, about three inches long, and appeared to be formed by the union of gelatinous bodies (swimming bells) of very complex form, and dissimilar at different parts of their length, so that the diameter of one-third was greater than that of the other two-thirds. I was much disappointed at my unsuccessful attempts to keep some record of them; but their invisibility, their fragility, and the approach of darkness, rendered all my attempts futile, and although I might have succeeded better if I had had another opportunity, I never saw anything like them on any subsequent occasion.

But the circumstance to be especially remarked is, that during all the time these curious animals were floating by, it was raining pretty hard—a condition which, *a priori*, might be supposed to have been most unfavourable for them; for the destructive character of fresh water to delicate marine animals is well known. Whence, too, could they have come in such profusion? And if the surface of the sea is their natural habitat, why are they not more frequently seen?

CHAPTER VII.

FORMOSA (*Continued*)—SAU-O BAY.

East Coast—Steep Island—Reefs at Sau-o—Chinese Village of Sau-o—Village of Tame Aborigines—Their Huts—Physical Characters—Dress—Native Cloth—Search after the Wild Aborigines—Characteristics of the Villagers—Their Occupations—An Alarm—They visit the Ship—Native Politeness—Language—Religious Ideas—Diseases—Distinctions from Chinese Race.

THE east coast of Formosa, as has been already observed, is remarkable for the absence of harbours; the mountains for the most part running sheer down into the sea. There is a landing-place at Chock-e-day, a considerable distance to the south;. but the only harbour is at Sau-o Bay, some 30 miles south of Ke-lung.

The Vice-Consul having called Captain Bullock's attention to some reported dangers about this important harbour of refuge, it was determined to visit it; for, although a plan of the harbour was appended to the chart of Formosa, since that plan had been constructed during a hasty visit, and represented only about six hours' work, it was probable that important improvements and corrections might be made. We had promised ourselves much gratification from a visit to this interesting locality, on account of its being a spot but very little known, and which very few have visited; and also because we hoped to see something of the aborigines of the island. Accompanied by the Vice-Consul at Tam-suy, and two or three other gentlemen, we accordingly left Ke-lung on the 12th June. After passing Petou promontory the contour of the coast changed, becoming less bold, and more retiring and flat, until we had reached the embouchure of the Kaleewan, one of the largest rivers in Formosa. This river flows into the sea, through a fertile plain 13 miles long and six broad, which supports about 10,000 inhabitants. Nearly opposite to it, at 10 or 11 miles distance, is a large island terminating in two peaks, the highest 1200 feet, the lower 800 feet high, presenting a precipitous face eastward to the sea; but, although we went on both sides of Steep Island, and near enough to see that it was cultivated in terraces to a considerable height on the landward side, we were unable to disembark upon it, and I cannot, therefore, speak with certainty of its formation.

The entrance to Sau-o Bay is protected or jeopardised (as the case may be) by a reef, which is nothing more nor less than a great trap dyke, running out nearly at right angles to the coast, and over which the waves dashed wildly, for the wind had risen. It extends a mile out, for the most part just above water, but rising into three prominent rocky peaks, one of which is 70 feet above low-water, and all three are whitened with the deposit of sea-birds which were resting upon them. Another reef, nearly at right angles to this, and probably of the same nature, runs across the harbour for about 300 yards, the highest point being a conical rock 15 feet high, the rest only just above water. It forms a natural breakwater, and, without blocking up the mouth of the harbour, shelters the interior, which is spacious, though not free from danger.

Sau-o Bay is shut in by lofty hills, for the most part steep, and densely clothed with forest. The formation is that of a compact, black, slatey rock, having a conspicuous cleavage varying in direction, and being in some places perpendicular to the level of the sea. There is no sandstone here, though there is abundance of sand upon the beaches. As we entered the harbour we observed, upon the north side, a hamlet of Chinese fishermen, consisting of half a dozen cottages on the hill-side, their boats being drawn up on the beach in front. Passing this by, we proceeded to the innermost or west side of the harbour, and anchored near a sandy beach, beyond which we could see the roofs of the houses of the principal village, called Sau-o, two other villages in the bay being, as we afterwards found, concealed from view. All the rest of the bay had a desolate and lifeless appearance, the wooded hills sweeping down to the water's edge, and presenting an aspect of wildness, which well accorded with our belief, that they were inhabited by the still savage aborigines of Formosa.

On the approach of the vessel, numbers of people assembled on the beach from the large Chinese village of Sau-o, attracted by the unusual circumstance: among them the *gamins* were conspicuous, capering about on the sand, while their more sober elders formed a long line in the background, squatting on their hams, and discussing over their pipes the cause of the phenomenon. As soon as we landed, we were escorted into the village by the crowd, and, on reaching it, were received by several explosions, which we were fain to consider a salute of honour. Sundry warlike-looking personages, armed with matchlocks, had turned out to meet the suspicious-looking strangers; but seeing us walking unarmed and amicably among the citizens, they fired their weapons harmlessly in the air for effect. They allowed us to examine their matchlocks, which we were told were manufactured at Amoy; and their ammunition, consisting of very coarse powder, with a finer grain for the priming, and bullets—some round, some oblong, some rectangular.

We found nothing remarkable in this village, which was essentially Chinese in its dirt, its pigs, and its inhabitants—closely resembling in character the other towns of Formosa; but our attention was arrested by a woman, whose handsome and European-looking features, and peculiar voice, at once marked her as non-Chinese, and showed her to be one of the aboriginal inhabitants. How she came thus domesticated among the Chinese we could not learn, but we heard from various quarters that a system of petty warfare is kept up between the two races, and that occasionally some of the women are carried off by the opposing parties.

The following morning we landed on the southern side of the bay, where we were to find the native village, of which no trace however was visible from the ship. We were met upon the beach by a number of men and women, who were in no respect, either of dress or feature, similar to the Chinese, and along with them, after the first expressions of surprise and curiosity, we entered the village. This is rudely walled, the entrance being through doors at either side, by which we passed into an assemblage of huts constructed chiefly of grass and bamboos. The grass is woven into a kind of trellis or mat, which is placed against the sides, while the chief part of the walls is constructed of upright sticks, the interstices being imperfectly plastered with mud to keep out the weather. The door is of bamboo, and fixed upon a rude hinge, the lower part revolving sometimes upon the bottom of an earthen cup, to give freedom of motion. The roof is a thick thatch of grass and herbs (in which a species of Turk's-cap lily is largely used), and is supported by bamboos irregularly disposed among it. At one end is often an overhanging shed, containing a supply of firewood, of which there is everywhere abundance.

The interior of these huts contained but little; a stone stove, and a square flat board in one corner, which did service for a bed—apparently for the whole family—appearing to constitute nearly the whole furniture. Articles employed in fishing might be seen stuck into the thatch, and a stool or two was to be found in most cottages. Besides these, a few small articles of convenience existed which could be found when required by the owner.

The occupants of these habitations were a fine race of people, much superior in good looks to the Chinese; their features being more regular and well-formed, and their expression decidedly more intelligent. The complexion was olive, the eyes wanting the obliquity so characteristic of the Mongol race, the cheek-bones less high and prominent, the lips somewhat thick, and the chin well turned, giving altogether a very pleasing expression, neither stupid nor savage. The hair was usually black, but sometimes had a decidedly reddish cast, and that of the women was luxuriant and tied with a loose knot, while the men had adopted the Chinese custom of shaving the forehead, though not so far back as the vertex, and wearing a pigtail. Their aspect and *physique* were in many cases very striking, and among them we saw both men and women of stalwart proportions. Some of the young girls were decidedly pretty, and exhibited all the coquetry, the love of finery, and other characteristics, which distinguish the sex in general in other parts of the world.

The costume of these people was somewhat slight. The

men were attired similarly to Chinese coolies, that is, usually in a simple pair of short drawers, to which, in some cases, a blouse was added. The dress of the women consisted of a short petticoat, folded round the loins and meeting in front, where it overlapped, but was not fastened. This petticoat did not reach so far as the knees, and the feet and legs were bare. A sort of loose jacket, open in front, completed their attire, though some of the matrons did not make use of this addition. This, however, appears to be a costume not always considered necessary, and those who landed at the village early on the second morning report that the population was more scantily clad, the men being entirely naked, and the women wearing only a flap round the loins. Seeing the strangers arrive, however, they retired with deliberation to their huts, closed the doors, and reappeared in the costume above described. The women possessed necklaces of beads, which they wore round their necks, and some of them had stone bangles round their arms; their ears were pierced in three or four places from the lobe upwards, though none of them seemed to have ornaments in them, except buttons, often of the commonest kind. The young children of both sexes were entirely naked. Most of them had objects round their necks, such as coins, beads, or buttons.

It should be mentioned that, in most cases, the garments worn by these people were made from a cloth of their own manufacture. This was a stout material, the threads of which were usually arranged in a zigzag pattern, and of a whitish or bluish-white colour. Many of the younger girls were employed in spinning the thread from fibres of hemp; and the cloth was woven by the older women, in pieces about a yard and a half long and a foot broad; three of which pieces they were willing to dispose of for one dollar.

These people are called by themselves *Kibalan*, and are, I believe, known by the Chinese as the *tame aborigines*, in contradistinction to the *raw savages* which dwell on the mountains, and on the east coast more particularly. These latter are at deadly enmity with the Chinese; while the Kibalans live in close proximity, though isolated from them. An officious half-caste among them informed us that there was another village close by; so, guided by him, we proceeded about a furlong along the beach, but were rather disgusted to find it a Chinese village, differing in no respect from other dirty Chinese villages.

Returning, therefore, to the Kibalan village, the name of which I believe is *Shek-fan*, we made known to them by signs that we were anxious to visit the mountains, and to meet with the savages. They, however, did their best to persuade us not to go, assuring us that we should be shot. On showing them our revolvers, however, they seemed to think we should be safer; but when we inquired for a guide, one and all declared that their throats would be cut if they ventured among the hills. After considerable parley, the sight of a dollar induced one to accompany us, and, when he had armed himself with his matchlock, we set out; our guide, however, taking good care to keep in the rear of the party. Crossing some padi fields, and proceeding along a sandy bay to the southward, our path was arrested by rocks, while on our right was a range of hills covered with a seemingly impenetrable forest. On closer examination, however, we discovered a beaten track, and up it we climbed, through a dense vegetation of tree ferns, camphor trees, etc., among which were some beautiful flowers, and many gay butterflies, although the overarching trees shut out most of the light. We ascended a considerable distance, tracing the path, which, although faint, was evident, and marking the trees as we ascended; but no trace could we find of the savages of whom we were in search. At length, finding the path less and less distinct, and time failing, we gave up the chase, and descended to the Kibalan village. The track which we had followed, however, was so evidently a more or less frequented one, that, considering the dread professed by the inhabitants of the village of their savage neighbours, it is difficult to understand by whom it could have been used, unless by the mountain aborigines descending to the plain in search of supplies. Subsequent inquiries at Sau-o, as well as some signs made by the Kibalans, elicited the information (whether true or false) that the aborigines of the mountains could not be reached under two days' journey from Sau-o Bay.

We unpacked our basket of provisions in the middle of the village, and were soon surrounded by the entire population, who pressed curiously about us, but withal civilly; and seemed to think our eatables not bad, particularly the loaf-sugar, which young and old appeared to appreciate. Captain Bullock had brought with him some old numbers of the *Illustrated London News*, which he distributed among them; but I remarked that though all seemed anxious to get a leaf, they did not look at the woodcuts, but immediately folded it up and put it in some part of their dress— nor could I interest them by pointing out to them the most striking illustrations, which they did not appear to comprehend. They did not, however, show any lack of interest and curiosity in most things, and the men particularly most inquisitively examined every part of our dress, feeling its texture, looking into our pockets, and showing by signs that they wished to see the interior of any box or bag we happened to carry. Nor were they content with looking once, but the same objects must be inspected again and again. The women more particularly exhibited a great anxiety to obtain as presents anything we could give them, particularly anything ornamental. The naval buttons were a great temptation; and over and over again they pointed to them, and intimated their desire to be the fortunate possessors of them. When denied, they would point to a young child and ask it for him, as though we could not then refuse it. Darwin, I may observe, makes the same remark when speaking of the Fuegians. If the button was given under these circumstances, it was immediately fastened on a string and tied round the child's forehead. But so importunate were they, that I might have completely stripped myself and found candidates for every article I possessed. Notwithstanding this, however, and that they repeatedly put their hands into our pockets, not a single article was lost, and no attempt was made to steal; but upon its being re-demanded, they never offered to retain any object whatever.

After our meal was over, the empty bottles were eagerly sought after, and we soon learned that no more acceptable present than *a bottle* could be made to them. This, which they called *brasco*, and *tobacco*,* were the only two things which they specially applied for. Nearly every one, men and women, smoked; and almost their first greeting was a demand for *tobacco*, a word which they appeared to have previously learned. A small plantation of tobacco grew and was in flower, within the walls of the village, and in

---

* Both words, no doubt, from the Spanish or Portuguese, *frasco* and *tabaco*.

several places the leaves were laid out in the sun to dry. It is smoked by them in pipes about a foot long, which the woman sticks in her hair when she is not using it, and not unfrequently we observed them smoking a bundle of scarcely dried leaves, rolled up and forming a rude and uncouth cigar.

It will not be supposed, however, that they were without occupation. In many huts the men were asleep, but towards evening they might be seen with their nets wending their way to the beach. Others I observed engaged over some seething vessels, in which I found they were extracting oil from the bones of turtle. The women had, several of them, naked babies hanging to their bare breasts; others came in from the country with pruning-knives, and laden with large bundles of grass and lily straw, which they laid down to dry in the sun, and which it appeared was ultimately intended for the repair of the thatch: these women had cloths wound round their legs, as a protection while in the field, and broad bamboo hats hanging by their side for wear in the hot sun. Others were spinning thread or weaving cloth, while some were engaged in beating rice out of the husks, which they did by placing it in a hollow stone vessel, under which was placed a mat, and then two of them beating it alternately with the end of a heavy bamboo.

While some were thus engaged, the idlers allowed themselves to be amused by some of our party, who showed them little tricks, which caused hearty laughter, and which they tried their best to imitate. Seeing a revolver, they were very anxious to have it fired off, and stuck up a leaf upon a door to be shot at, which was done twice, upon which there immediately appeared two or three men armed with matchlocks, who had evidently turned out at the sound of the pistol to protect the community in case of need. This little incident seemed to prove that they were always on the alert, and gave colour to the general report that they, like the Chinese, are subject to the raids of the raw mountain savages, against whom they are always more or less prepared to defend themselves. This was also corroborated by the fact that in the midst of the village a building was in course of construction which was evidently of a defensive character. It was in a very unfinished state, having at present no roof, and the walls not all completed; but the loopholes in the walls of the finished sides, as well as the accounts of the natives themselves, showed for what purpose it was intended. Men were engaged in sawing wood, and doing other business of construction.

By the side of this unfinished building was piled a great heap of tiles of a dark colour, and of a most rubbishy brittle character, which they had purchased of the Chinese for roofing their fort. It seemed the greater pity, inasmuch as the rocks of Sau-o Bay are of a slatey character, with very distinct cleavage; and, close by the village, slates might have been obtained by a little trouble and intelligence; and these would have answered the purpose better than the wretched tiles they were about to use. A herd of water-buffaloes, brought home late in the afternoon, repaired to a muddy pool in front of this building, and, with their characteristic timidity and stupidity, after eyeing us curiously for some time, they took alarm at some movement of one of our party, and bolting helter-skelter out of the mud, floundered over the heap of tiles, crushing numbers of them to pieces, and all but overturned some of the native huts in their mad career. Besides buffaloes, they have pigs (always black), Chinese dogs, Malay cats with short twisted tails, and fowls.

The following morning a number of natives visited the ship in their boats, and, on bottles being shown to them, they eagerly demanded them. When thrown into the water, half-a-dozen men leaped after them, and vigorous swimming-matches took place for the prizes. They would also dive for buttons of any kind that were thrown in. Soon afterwards several boats, full of people of all ages and both sexes, came alongside and readily ventured on board. Indeed, one great distinction between these people and the Chinese was the entire absence of timidity on the part of the females, who, instead of running away and hiding themselves as soon as they are even looked at, showed the most perfect confidence and freedom from *mauvaise honte*. They immediately commenced eagerly inquiring for *brascos* (bottles), and as our supply of these desirable articles was limited (owing to the custom of throwing empty bottles overboard), considerable jealousy was excited among the unsuccessful competitors. There was no idea of barter, and perhaps it was our own fault that we obtained nothing in return for our valuable presents. A number of our visitors were induced to descend to the captain's cabin, where, as lunch was going on, they readily partook of the edibles, and made themselves quite at home. Captain Bullock good-naturedly cut off the tassels from his cushions, which were immediately transferred to the hair of the native beauties. After going about the ship, and conducting themselves with the greatest propriety, they returned to the village. A little incident struck me as worth recording. One of the men passing the ward-room sky-light, where some of the officers were at lunch, looked down, and lingered, when he was pulled away gently by another man who was with him. It was a slight movement, but *Chinamen* would have remained and stared till their eyes started from their heads before such native politeness would have occurred to them.

Mr. Sutton, chief engineer, took his camera on shore, and succeeded in taking several excellent stereoscopic pictures of the village and its inhabitants. The people readily acceded to the desire that they should sit, and several picturesque groups were formed, some of which were successfully fixed by the camera: of course it was very difficult to keep them all quiet, and impossible to make them comprehend the necessity of absolute stillness during the critical moments. The result, however, was in several instances very satisfactory.

The attempt to learn some of their words, and to form as good a vocabulary * of their language as the time would permit, was met by perfect good will on their part; and many words, as well as their mode of counting, were obtained, chiefly from the women, who appeared to take considerable interest in imparting the information. Considerable amusement, too, was excited by our mistakes in pronunciation, etc., and our efforts elicited a considerable show of intelligence on their part. We found the women much more serviceable than the men for this purpose, chiefly on account of their clear pronunciation, which was decidedly more distinct than that of the men. In all cases, the attempt of the women to pronounce English words was more successful than that of the men. The word "flint," for instance, being given them, a man would not approach it nearer than *plin-iss*, while the women at once said *fil-lint*. The voice of the women was remarkably agreeable, having a plaintiveness and softness which

* This vocabulary will be found in an Appendix at the end of the volume.

were really striking, and sometimes sounded more like a gentle singing than speaking.

We looked in vain for any indication of their religious ideas. Over the door of the village, by which we entered, some one had stuck a joss-paper, after the manner of the Chinese, and probably some Chinese had done it, but there were no joss-houses or temples in the hamlet, nor did we find any in the houses, though among the poorer Chinese almost every house has a little altar to the lares in the principal room. We inquired as well as we were able of the inhabitants on this point, but could elicit nothing from them. Nor could we discover any indication of a written language.

With regard to their diseases, we had no means of learning anything of the mortality of the village. One young woman appeared to be recovering from small-pox, and one old woman was covered with a skin disease, which gave her a leprous appearance, but the people in general were healthy-looking and physically strong, hardy, and well-made. We observed no deformities among them, with the exception of one child of three or four years old, which crawled nimbly about on its hands and knees, but appeared to be physically unable to stand or walk. The village may have contained, at a rough estimate, 250 inhabitants. There were plenty of children, but old, grey-headed persons were not numerous.

In conclusion, I think it is evident that the race of people inhabiting this village is distinct from the Chinese. Among the women, particularly, there was scarcely one who had a Chinese feature, and their habits and modes of life also differ considerably. The feet were in no instance bandaged, as is universally the custom among the neighbouring Chinese. With regard to the men, it was not always so easy to discriminate, although in many, or rather most, instances, the Kibalan man was bigger and more stalwart, and with a cast of features superior to that of the Chinese. Some of them may have been half-castes; but I am of opinion that the majority of the inhabitants were of pure aboriginal descent, though how they became separated from the mountain savages, and the process and reason of their domestication, I have no means of knowing. Their present isolation in their own special village in a great measure accounts for the apparent purity of their blood. They were in all respects a more intelligent and more engaging people than the Chinese of Formosa, though these latter affect superiority. Thus, when I inquired of a man in the Chinese village of Sau-o, who I imagined had a dash of Kibalan in his face, if he belonged to that race, he replied, " No, I am a *man;* " (that is, a *Chinese*, not a foreigner).

On the third day we weighed anchor and stood out of the harbour; but we had scarcely got in motion when a bump upon a sunken rock warned us that the dangers of Sau-o Bay were not yet fully known. The anchor was at once dropped, and a search made for the rock, which was at length discovered 12 feet below the surface; but as we, fortunately, only drew $12\frac{1}{2}$ feet water, and had but little way on, no damage was done. Had it happened on our entrance it might have been more serious. This circumstance gave me, however, more time to inquire into the peculiarities of the natives of this interesting and little known spot.

CHAPTER VIII.

THE ISLANDS NORTH-EAST OF FORMOSA.

Visit of a Chinese Admiral—Ke-lung Island—The Harbour from the Sea—Pinnacle Island—Craig Island—The Wideawakes; their Breeding Place—Geological Structure of Craig Island—Hunt on the Rocks—Grapsi—Agincourt Island—Pinnacle Rocks—Hoa-pin-san and Tia-usu—The Raleigh Rock—The Dredge—Chromodoris—Gigantic Foraminifera—Further Search—Return to Ke-lung.

ON our return from our visit to Sau-o Bay we found that a Chinese admiral was in the harbour of Ke-lung, his flag-ship being a gun-boat of British build. An interchange of civilities took place, and the admiral paid us a visit on board the "Serpent," accompanied by his interpreter, a young man who had been engaged by the English at Canton, and at once recognised a portrait of Captain Bate (who was killed before Canton) which was hanging in the cabin. The admiral was a jolly, though spare, elderly man, very pleasant and affable, and at the same time extremely inquisitive with regard to everything which he saw on board the English ship. He was never tired of asking questions, through his interpreter, about the fittings and tackle of the "Serpent;" but was, apparently, more particularly interested in the Armstrong guns, which he examined with great care, and the manipulation of which he watched attentively. He seemed greatly to covet our two small 20-pounders, and inquired their value. Being told that they cost 1500 dollars, or

120*l.* each, he at once offered to give that sum, and seemed disappointed when he was informed that Government property could not be disposed of. Having lunched with us in the cabin he returned to his ship, with many polite *chin-chins*, receiving a salute of three guns as he left the "Serpent." In the afternoon he quitted Ke-lung, saluting us with three guns as he passed, having previously sent on board a present of sweetmeats, and his card on red paper.

Whilst in this part of Formosa it was determined to search for, and determine the position of, the Raleigh Rock and Recruit Island, both of which were very doubtfully laid down on the charts. The Raleigh Rock was supposed to be seen by H.M.S. "Raleigh" in 1837, and was afterwards described by Sir E. Belcher as 90 feet high, perpendicular on all sides, and covering an area of about 60 feet in diameter. The ship "Recruit," in 1861, sighted an island near the same spot, which was described by the master as about a mile in extent, and 600 feet high. The ship "King Lear," Captain Croudace, also describes a rock of 90 feet high close to the same locality, rising very abruptly, and having a small rock standing erect, like a pillar in ruins, detached from its north side. The possible existence, therefore, of two or three large rocks in the track of vessels, in a locality in which the reckoning was much affected by the Japan stream, was of sufficient importance to warrant an attempt to clear up the mystery, especially as Admiral Belcher had not professed to settle its position, owing to bad weather.

On the 1st June, therefore, we quitted Ke-lung harbour on the quest, for the second time; for on the first occasion, four days previously, we encountered such heavy weather on leaving our shelter, that we were glad to put back with all speed, and had received no encouragement to quit it again until now. Immediately outside the harbour, Ke-lung Island is a very striking object, and a fine landmark for the entrance, from which it is distant only two and a quarter miles. It is a steep conical rock, rising 580 feet above the level of the sea. I had no opportunity of landing upon it, and cannot, therefore, speak with certainty as to its geological structure; but from its peculiar form, and from the fact that between it and the mainland there is everywhere 30 to 35 fathoms water, I should suppose it to be probably volcanic. I more than once passed within a mile of it, and could see no signs of sandstone; nor are there any trees upon it, as there are upon the sandstone shores of the harbour.

Looking back towards Ke-lung, one could not fail to be struck with the appearance of the coast northward of the harbour, which consists of a series of ascending strata, dipping south at an angle of 15°, and hollowed at the outcrop northwards with valleys extending as far as the eye could reach. The conformity of the lines is very remarkable, and they are continued at the back of the harbour in the distance; but it cannot be observed on the southern side.

Immediately north of Ke-lung we met with a group of three islands—Pinnacle, Craig, and Agincourt—little, if ever, visited, and of which no description has been given. The first of these, Pinnacle Island, is of a remarkable form, and has received the native name of the *Chair-bearers*, from the fact of the outline faintly resembling a Chinese sedan borne between two men. It is a perfectly bare craggy rock, with a tall pinnacle at either end, against which the waves dash furiously, sending the spray a hundred feet high. The rock was whitened with the excrements of sea-birds, and I had no opportunity of a close inspection.

On approaching Craig Island great flocks of birds flew

about, making a great noise; and large white patches upon the hill-side proved, with the aid of a telescope, to be unquiet flocks of gulls and tern, in constant movement. On searching for a landing-place we observed, somewhat to our surprise, a rude hut, with a piece of blue cloth waving before it and doing duty for a flag. Seeing one or two human figures, we at first took them for shipwrecked mariners; but we soon discovered that they did not seem anxious to be relieved from their position, and they ultimately proved to be two Chinese egg-collectors. In their huts were large numbers of eggs, but for what purpose they collected them, how they came upon the island, or how they were to get off (for they had no kind of boat) was a puzzle, and remained so, for our Chinese servants could hold no intelligible communication with them.

A tent was pitched on the beach for the purpose of taking sidereal observations; but I preferred remaining on board till daylight. The noise and chattering of the birds could be heard all night, one now and then crossing the ship; and at dawn I landed and walked through the thick herbage, laden with dew, to the top of the island. Every here and there, in a clear patch, a number of wideawakes, differing in no respect from the wideawakes of Ascension, were seated, scattered at intervals over the ground. On my approach they chattered and croaked, and made as though they were ready to run at me; but thinking better of it, they would rise with a clumsy fluttering, and take wing. They mostly perched upon the large overgrown stones, from which they rose easily; but if they happened to be upon the herbage they floundered along, vainly endeavouring to rise, until they reached the edge of a stone, over which they tumbled, giving their long wings room for extension. They were pretty birds, black above and white beneath, face white, beak and feet black, and tail forked. There were vast numbers of them, and it would have been easy to have taken with the hand as many as one wished. On rising, they formed a thick canopy immediately overhead, darting at our hats, and almost at our faces; so that we were under the necessity of holding up sticks and waving them over our heads to keep them off while we stooped to pick up some eggs. All this while they made various noises, chattering, croaking, and barking like a dog.

The nests of these birds were mere depressions in the ground upon the hill-side; but some of them chose the rocks, or crept into little clefts which were only just large enough to admit them. A great many of them had eggs, but I nowhere found more than one egg in a nest. These eggs were very variously marked, sometimes brown, speckled with greenish, in colour like that of a magpie; sometimes uniformly speckled with small brownish spots upon a white ground; while others again had larger blotches about the big end. They appeared to have been systematically taken from them, and many of the birds were sitting upon a small rough piece of rock.

Besides the wideawakes, there was a large number of birds of another species, somewhat larger in size and of a blue-grey and white colour, and these formed the large patches seen upon the hill-side; but they were wilder than the terns, and all flew off on our approach. These birds had also eggs, of a larger size than those of the terns, and blotched with reddish-black on a white ground.

Besides these there was a small sooty petrel, and a few gannets (Sula alba); and the only bird I observed upon the island which was not aquatic was a tree-sparrow (Passer montanus). On the rocks by the shore were a number of dove-coloured birds with white foreheads, of which, however, I failed to obtain a specimen; nor could I find their nests and eggs, unless a white egg, like that of a pigeon, which I found in a crevice, were one of theirs.

The whole of Craig Island is a mass of trachytic lava, broken up into smallish rough masses, even to the very summit. These being more or less covered with grass, rendered walking over it very difficult. The blocks upon the sea-shore are very large, and piled up in picturesque confusion. On the eastern side is a series of magnificent lofty pinnacles, or aiguilles, perforated below, and thus forming beautiful natural arches, which are grand and imposing objects seen from the beach. These pinnacles appear to be portions of a trap-dyke running out into the sea. The surface of the island is somewhat disintegrated, and a poor soil is formed, upon which a large number of herbs are growing; but there are no shrubs or trees. The herbs consist principally of a succulent Saxifrage and a species of sea-cabbage (Brassica), with pink flowers; but there is no inconsiderable variety of vegetation.

On a sandy part of the shore, where the tent was pitched, we found a number of large centipedes by scraping up the sand and lifting up the stones; there were also ants, and a few hemipterous insects, and cockroaches (Blattæ). I also found in other parts of the island green beetles (Euchlora), like our rose-beetles. A ramble among the great blocks of trachyte strewing the beach did not yield much, although the rock-pools formed exquisite aquaria, in which were tunicates, sea-anemones bearing a close resemblance to our Actinia bellis, rock-fish, and some sea-weeds which seemed of a brilliant blue while under water, but when taken out lost all their colour. Fine specimens of the barnacle, Pollicipes mitella, were wedged in clefts of the rocks; and running about hither and thither upon them were numerous active crabs, of the genus Grapsus, which are more or less characteristic of all tropical islands. They are somewhat quadrilateral in form, and have a flattened aspect, and are so wary that it is a most difficult task to capture them, more particularly as they are always found upon irregular, and often smooth rocks, over which they run with great velocity, in places where it is impossible for their pursuers to find a footing. The first pair of legs is short, curved, and spinous the other pairs usually more or less compressed, and hairy to the extremity. With these they hold securely to the most slippery surfaces. The most generally distributed species is one (G. strigosus) ornamented with long wavy lines of red and orange, and a very beautifully marked animal. Their cast shells of all sizes strew the rocks in all directions, and are scarcely distinguishable from the living animals, except that, perhaps, the tints are less vivid than those of the living ones. When taken up, however, they fall to pieces, unless great care is used. It was amusing to watch the Grapsi, which always seemed to know when they were at a safe distance, and then did not trouble themselves to move; but stir a foot, and they would scuttle away in all directions, and, if closely pursued, escape into crevices, or down gullies between the rocks, where it was impossible to follow them. I have seen these crabs not only move up and down nearly perpendicular surfaces, when undisturbed, but leap sideways over crevices several inches wide, a feat which they performed with singularly little effort, and so rapidly as scarcely to attract observation.

But my time was too limited to allow of my doing more

than take a cursory view, and note the characteristic productions of the shore, when I was summoned to rejoin the ship, or be left, Robinson Crusoe-like, upon the island. My towing-net from the ship, however, meantime yielded a number of the little oceanic crabs with spiked carapace (Lupea pelagica).

Agincourt, the third island of this group, presents a remarkable appearance from the numerous caves in its sides, which are visible from a considerable distance, and its structure on a near inspection is easily discernible. The island is formed of a rounded hill of sandstone, with several smoothly worn eminences, and traversed from end to end by an enormous dyke of trappean rock. This dyke, best seen on the north side, is broad and nearly level, terminating in an abrupt precipice on the left, and gently sloping towards the sea on the right. It cuts off a small portion of the sandstone rock from the main mass, and in this portion are two conspicuous caves, finely and spaciously arched at the entrance, but apparently not penetrating very far back. There are no less than six caverns in the sides of this island; nor are they all confined to the soft sandstone, for while two of them are in the sandstone of the north side, and two in the sandstone of the south side, the remaining two are situated in the abrupt face of the trap cliff on the eastern side of the island. In all the sandstone caverns the arches were broad, sweeping, and symmetrical; but in those of the trap they were lofty and irregular in form, and quite distinct in character from the rest.

On the west side is a poor village, or hamlet, whose inhabitants we could see watching us; and this accounted for the fact that the highest part of the island was under cultivation, and also for the absence of the numerous birds which characterised the neighbouring Craig Island. Agincourt was covered more or less with vegetation; but there were no trees, and only a poor apology for shrubs. The sea in the immediate neighbourhood of this group has a general depth of rather more than 100 fathoms.

About 75 miles to the E.N.E. of Agincourt is the second group of islands, consisting of Hoa-pin-san, Tia-usu, and the Pinnacle Rocks, the last consisting of several distinct islets, and forming, with Hoa-pin-san, one group. Hoa-pin-san is composed of trappean rocks, with a bold outline, and rising nearly 1200 feet above the sea; while the Pinnacle Rocks well deserve their name, from the remarkable forms which the most elevated and prominent of them assume, and which look like buildings, lighthouses, &c. I did not approach them near enough to ascertain their structure; but Sir Edward Belcher * says, with great probability, that they are masses of grey columnar basalt, upheaved and subsequently ruptured, and rising suddenly into needle-shaped pinnacles, which are apparently ready for disintegration by the first disturbing cause, either gales of wind or earthquakes. Tia-usu also is composed of huge boulders of a greenish porphyritic stone, probably a basalt, cemented by coralline and amygdaloidal matter.

On none of the islands of this group are there any trees or shrubs—they have a rocky and desolate appearance, only relieved by the multitude of birds which darkened the waters around by their vast flocks, as they sought their resting-places towards sunset. The islands, too, are more or less whitened by their deposits. The soundings around this group reach 80 fathoms.

In the night we sighted at eight miles distance a rock,

* Voyage of the Samarang, vol. i. near the end.

which seemed to be that of which we were in search; and, steaming up, we lay off near it till morning. It then appeared as a large irregular precipitous rock, about 600 feet long, and rising 280 feet high, with a reef at either end—covered with sea-birds, and whitened with guano, which ran down over the ledges in long streaming festoons, giving the rock a very singular appearance. Great numbers of gannet (Sula alba) were flying around, and it appeared that they were the principal inhabitants.

The sea was unfortunately not smooth enough to enable us to effect a landing; but as the ship was drifting a short distance off to allow of taking observations, I put down the dredge in about 60 fathoms. It came up richly laden with sponges, delicate branching corals, and Gorgoniæ of the richest colours—yellow, red, green, brown, &c.—zoophytes, tunicates, small shells, Ophiuræ, &c.—a very *embarras de richesses;* but the most interesting haul I had ever seen. There were several pretty little feather-stars (Comatulæ), a little nymphon, or sea-spider, and, within the sponges, small crabs had their habitations.* It took a long time to search through the contents of the dredge; and so numerous were the species of animals, that, unassisted as I was, it was perfectly vain to think of doing more than select the most interesting for examination. In the midst of this mass of coral *débris*, I found a magnificent Nudibranch of a new species, but probably belonging to the genus Chromodoris. It was nearly three inches long, translucent, of a deep amethystine tint about the head, shading into reddish upon the back; all round the mantle was an edging of opaque white, while the laminated tentacles and leaf-like gills on the back

---

* These crabs constitute a new genus, which Mr. Spence Bate has named Spongæcetor.

were of a rich orange-yellow colour. A more splendid animal I had never seen, even among this surpassingly beautiful family; and notwithstanding the unenviable position from which I rescued it, it became quite lively when placed by itself in sea water. It was active and graceful, and lived several days, giving me time to describe and figure it; after which, I gave the little creature an honourable position in a series of bottles which contained my scientific novelties.

At the same spot, there came up in the dredge some curious round bodies, which might have been taken for small oyster-shells, but which were in reality of far greater interest. These were members of the family of Foraminifera—lowly organized creatures, for the most part microscopic, but which in this case were possessed of a very definite size. The greater part of these were Orbitolites,—round, button-like, flattened shells, of a most remarkable symmetrical structure, which has been admirably elucidated by Dr. Carpenter. They were $\frac{1}{3}$ inch in diameter for the most part, and quite white. They consist of a concentric series of alternating cells or chambers, directly communicating laterally, and indirectly in a radiating manner. This regularly-constructed calcareous framework is built up by, and is filled with, a sarcodic substance which sends out stolons through all the canals, and ultimately passes out in filaments through a series of pores along the margin, which filaments have the functions of prehension and locomotion combined.

But besides these comparatively well-known forms of Foraminifera, there were others which are at the same time among the rarest, and are the giants of their tribe. These belong to the genus Cycloclypeus, and became known by some specimens dredged by Sir E. Belcher off the coast of Borneo. These specimens measured in some instances $2\frac{1}{4}$ inches in diameter, and some of them are in the British Museum. It is interesting to observe the spread of this form fully 20° further north, in the N.W. corner of the Pacific —for one of these great Foraminifera, $1\frac{1}{3}$ inch in diameter, I found off the Raleigh Rock. Its edges were somewhat broken, by being crushed up in the dredge; but what was very remarkable was that it bore evident marks of having undergone severe fracture in an earlier stage of its existence, which it had patched up, and in process of growth had become quite round and orbicular once more. In Cycloclypeus there is a great advance upon the simple structure of Orbitolites, though in outward appearance they closely resemble one another; but the cells which enter into the construction of Cycloclypeus, instead of being closely connected with one another, are singularly isolated, while the shelly covering is wonderfully elaborate, as may be seen by referring to Dr. Carpenter's figures in the Philosophical Transactions (1856).

Notwithstanding that I had already more material than I could possibly manage, I was anxious to accumulate more riches, and hoped to make more discoveries, and the dredge was again dropped overboard, while the examination of the first mass was continued. At length it was drawn up, and full of expectation I looked over the side, but alas! the handle alone came to the surface, with one scraping side still attached, but bent nearly double. It had caught against a rock, and unfortunately not being a dredge of the self-disengaging kind, a lurch of the ship had torn it in halves, and left frame, net, and contents at the bottom of the sea. I had no other dredge at hand, and it was impossible to replace it then and there; and had therefore for the present to content myself with the towing-net, which yielded a number of minute banded fishes, Medusæ, Sagittæ, and the Pteropods Cleodora and Hyalæa.

Having fixed the position of the Raleigh Rock, we then went in search of Recruit Island, or any other similar object which could have given rise to the descriptions of a rock 90 feet high, and standing on a base of 60 feet, looking from a distance "like a junk under sail." But although we went ten miles further east, and searched carefully over the position assigned in the Admiralty charts—Belcher's, the Raleigh's, Croudace's, Horsburgh's, and Lyall's—we failed in finding any sign of an island or rock; and the conclusion was forced upon us that the island I have described was at once the Raleigh Rock and Recruit Island; and this object, whose very existence had become somewhat mythical, had now taken, and would henceforth keep, a definite and authentic position upon the map of the North Pacific.

An uniform depth of 70 to 80 fathoms existed everywhere around the Raleigh Rock, and the lead constantly brought up sand, foraminifera, shells and echinus' spines. The object of this excursion being now accomplished, we returned once more to Ke-lung harbour; but only to make preparations for finally quitting it for the Chinese coast.

CHAPTER IX.

HAITAN STRAITS AND COAST OF CHINA.

Red Discoloration of the Sea—Haitan Island and Straits—Middle Island—New Anemone—Black Islet—Its Fauna and Flora—Chinese Pirates—Rumbling Fish—Slut Island—New Nudibranchs—Iridescent Seaweed—Trigger-Shrimp — Comatula—The River Min — Pagoda Anchorage — Chinese Pagodas—Shwin-gan Passage—Luminous Sea—Plague of Flies—Insects at Sea—Wosung River—Shanghai.

On June 22nd we crossed the Formosa channel for Haitan Straits on the Chinese coast. The only noticeable occurrence on the passage was a remarkable discoloration of the water—for once of a red colour. The sea swarmed with vast numbers of small gelatinous worms, about half an inch in length, and of a pinkish colour, which accumulated under the sides of the ship in immense quantities, as she drifted for current in mid-channel. The towing-net brought them up in solid masses of red jelly, and when placed in a basin they swam about with great activity and with a vermicular movement. This was, with one exception, the only occasion on which I observed a red discoloration of the sea. In the other case, when between Aden and Galle, the water appeared in many places to be alive with myriads of minute crustacea, which formed reddish patches; and when the sun shone upon the sea they could be discerned from the ship's side darting rapidly about below the surface.

The following morning at daybreak we sighted Haitan Island, and coasted along it for some time southwards.

This island, as well as the adjacent coast, presented a most barren and desolate aspect, and consisted of perfectly bare rocks, apparently of whinstone, descending smoothly to the water's edge, interspersed here and there with sandy patches and conical hills of sand. A few houses were visible, and attempts at terraced cultivation, but no trees or shrubs were anywhere to be seen. The coast of China is for the most part no less uninviting than that here described, being usually wild, rugged, and barren.

Entering the Straits at the south end we passed through a maze of small islets, between Haitan Island and the mainland, which all partook more or less of the same character.

Two of these small islands, which I had an opportunity of exploring, were types of the rest. The first of them, called Middle Island, was formed of rugged granite rocks promiscuously heaped together, the upper part being much disintegrated and formed into a barren soil. A pair of oyster-catchers (Hæmatopus) inhabited the island, and were very tame. Two species of Coccinella and Epilachnia (ladybirds), one of them identical with our common seven-spotted ladybird, and some small Hemiptera were all the insects which rewarded my search, and a broken Helix of small size alone represented the land-shells. The rocks, even above high-water mark, were thickly covered with a small and very prickly oyster (Ostræa spinosa), and in the clefts grew the large barnacle, Pollicipes mitella. A few Purpuræ seemed the only living mollusks; and in a sandy bay a number of small and exceedingly swift-running crabs (Ocypoda) scudded along, and eluded pursuit in the most provoking manner, suddenly changing their course and running the opposite way, *without turning*, and then darting like lightning into a hole in the sand, which they had already prepared. Besides these there was nothing more to be noticed, with the exception of a very elegant anemone (Sagartia), of a new species, which I afterwards found to be extensively distributed over the islands in these Straits. It was of a small size, adhering firmly to rough stones, and very prettily marked with a close vertical series of hair-like streaks, which, when the animal was expanded, ran parallel from mouth to base. These lines are variable, and give considerable variety of appearance to the individual zoophytes. The greater number, and usually the broader lines, were olive-green, or greenish brown, and this may be called the body-colour, and these are alternated with dull white, and, at regular intervals, a streak of vermilion or yellow, perhaps twelve vermilion streaks altogether. In some specimens the white streaks were replaced by yellow, while in others the colour was a uniform olive, with about a dozen delicate hair-like streaks of vermilion or orange placed at regular intervals.

The other island, called Black Islet, was a small spot, about a furlong in length, consisting of a mass of granite broken up and much disintegrated—all the upper part being a soil of coarse quartz sand. It was within sight of the last, but the rocks were perfectly smooth (except for a few small barnacles), and had not a single prickly oyster growing upon them. Here, however, were abundance of Ligiæ, but the tide was too high to allow of my exploring low down. I therefore devoted the hour at my disposal to the higher parts of the islet; and considering its small size, the variety of its Fauna and Flora was remarkable. The Fauna was composed chiefly of insects—three species of dragon-fly (Libellula), a grasshopper, a small spider of the genus

Acrosoma, two or three species of Microlepidoptera, some two-winged flies, ladybirds (red with black spots), a little hemipterous insect, and a small beetle (Opatrum), whose habitat was on the sandy beach. There were also some terns flying about the island.

The Flora consisted of 32 plants, viz. three Cryptogams (a fern, and two lichens), and 29 Phanerogamous plants, of which 16 were in flower, and belonged to the following genera:—Rumex, Oxalis, Rubus, Statice, Aster, Atriplex, Saxifraga, Sonchus, Silene, Arenaria, Vicia, Lathyrus, Allium, and Mesembryanthemum. Besides these, there was a thistle (Cnicus), a dwarf acacia, a Juncus, a grass (Poa), a Soldanella, a crucifer with long pod, and a small suffruticose Juniper.

All the islands in Haitan Straits are of volcanic structure, and present several picturesque points—as the Pillar Rock, a lofty rectangular mass of stone of imposing appearance, and Junk-Sail Island, which at a distance well bears out its name.

We were detained in these Straits for about a week by bad weather, which amounted to a typhoon outside. During this period numerous junks sailed through, some of them having a very suspicious appearance, and such as a becalmed merchantman would hardly like to meet; for notwithstanding the active services of our gunboats in the suppression of piracy, "pylong pidgin" flourishes when occasion offers. So recently as May, 1865, the "Formby" British ship, becalmed off Cape Padaran, was surrounded by 15 piratical junks, which suddenly hoisting a red flag, made an attack which would probably have proved successful, had not a fortunate breeze sprung up, which enabled the "Formby," after defending itself for some time, to get away from its persecutors, who then reluctantly gave up the attack. About the same time, the bark "Ruby" was attacked by pirates, also near the above locality, and having no means of defending themselves, the captain, with his daughter, who was on board, and the crew, got into their boats on one side of the ship at the same moment that the pirates boarded on the other side. Fortunately they were not regarded, the miscreants betaking themselves to plundering the ship, which they did most effectually, tearing up the boards and otherwise damaging her, while the crew were left unmolested. The pirates afterwards abandoned the vessel, which was found derelict by a Hamburg barque, and taken into Saigon. Meanwhile the captain and crew, after having been several days in the boat, were picked up by a French gunboat, and taken also into Saigon, where the first object that met their eyes was the lost vessel safe in the harbour. Wreckers, too, turn up in a wonderfully short time, when there is anything to plunder; and some months subsequently to our visit, a British barque, the "Fanny," having been abandoned at the north entrance of these Straits, was boarded and looted; while the crew, who had taken refuge on Slut Island which we had found deserted, were then set upon, and robbed of the little they had managed to save.

In order to protect shipping, and to assist in putting down Chinese piracy with a strong hand, it was, not long since, announced that certain new gun-boats were to be built to reinforce the China squadron, and a number of officers were to be sent out. It is to be doubted, however, whether our active interference with an evil whose root lay in the Government of China, is entirely for good. No one who knows China can doubt that the crying mischief lies at the doors of the mismanagement and indolence of the Court of Pekin. It is their interest as much as ours that piracy should cease; nor are we alone among European powers in suffering inconvenience from the outrages of piratical ships. Of course the Chinese are willing enough to be freed from embarrassment by our kind offices; and the French, Germans, and Americans, whose ships also throng those seas, would no less benefit by its suppression. But it has been suggested, with show of reason, that it would be suicidal to our own interests if we unconditionally render aid to the Chinese Government, and thus take away their only motive for action, while, at the same time, the expense of fitting out aggressive gun-boats ought to fall, not only upon us, but also in fair proportion upon those who will reap some proportion of the advantage.

When lying at the north entrance of Haitan Straits, I heard one evening a singular sound, which was attributed to a fish—whether rightly or wrongly I had not the means of ascertaining. It was a loud rumbling noise, which at one time might have been taken for the distant roar of the sea, at another, for the singing of a kettle. When once attention was called to it, it appeared very distinct and loud, but it was of that monotonous nature that it might have remained unnoticed for a considerable time. It was most observable on the starboard side of the ship, and in the cabin. The scuttles were closed, however, and on deck I failed to hear it. I listened long to it, hoping to find some explanation; but could only suppose it might have been produced by some fish, or other animal, under, or against the side of the ship. The sound became louder and fainter at intervals, and finally it ceased by imperceptible degrees.

Between us and the mainland was a small island (Slut Island), which I visited several times, and found productive of many interesting species. The island itself appeared from the ship to be highly cultivated, a series of green terraces extending nearly along the whole length; but a nearer approach showed the terraces to be covered with rank grass, and at the foot of them a ruined village, the bare walls of which were thickly overgrown with coarse vegetation. Two or three men, who were employing themselves in picking up what edible articles they could find on the beach, assured me that the inhabitants had voluntarily deserted it; but it had all the appearance of a pirate village, which had been destroyed by fire and sword. And, indeed, its situation was most favourable for such lawless pursuits. The men had taken possession of a joss-house, which was in pretty good preservation, and contained, among other figures, some of a frightfully diabolical character, which, however, did not appear to deter the Chinamen from cooking their shell-fish in their immediate presence, and under their influence.

Upon the stony shore at low water were shells of the genera Turbo, Ranella, Murex, Oliva, &c. Large purple Echini, and a few inconspicuous star-fishes, represented the Echinoderms; and the little striped Sagartia before described was here pretty numerous. Among my trophies also was a Virgularia about six inches long, closely resembling V. mirabilis. On one shore-excursion I was so fortunate as to find four new species of Nudibranchiata. The first of these was a Doris with a rich chrome border, and large, circular, raised carmine spots scattered over the back, giving it a very striking appearance. It was a small animal, and I found but one specimen, and this makes it the more remarkable that I should have rediscovered the species afterwards on the shores of Labuan. The second species was also a Doris, twice the size of the last, bright yellow bordered with mauve. The third was a Doridopsis, nearly two inches in length, of a rich velvety brown, with capacious branchial leaves upon the back; and the fourth was a small and elegant species, probably of Plocamophorus, but, like the other three, undescribed.

A very beautiful seaweed, a species of Valonia, grew in the rock-pools of this island. It was small in size, and, seen beneath the water, had a most splendid glossy green appearance, which might with truth be described as luminous, strongly calling to mind the extraordinary moss, Schistostegia, which is found in the caves of North Lancashire. This moss when brought out of the cave loses its brilliant appearance, and similarly the Slut Island seaweed entirely lost its peculiar golden aspect when taken out of the water.

Among other animals which I brought from the shore was a small shrimp, a new species of the genus Alpheus, of a deep violet colour, and with a claw of very remarkable construction. I placed it in a basin of water with a small crab, whose presence appeared violently to offend it. Whenever the crab came in contact with the shrimp, the latter produced a loud sound, as though some one had given the basin a sharp tap with the finger-nail. During the night we were frequently startled by this sound, the explanation of which was as follows. The shrimp possessed two chelæ or claws—one, the right, a large and stout one, and the left one long and slender. When irritated it opened the pincers of the large claw very wide, and then suddenly closed them with a startling jerk. When the claw was in contact with the bottom of the basin, a sound was produced as though the basin were smartly struck; but when the claw was elevated in the water, the sound was like a snap of the finger, and the water was splashed in my face. The appearance of the claw during the operation reminded one forcibly of a trigger. The pincers opened slowly, till they gaped very widely, and when they had opened to their fullest extent, they closed instantaneously, as though with a spring, like the trigger of a pistol. If the pincers were not opened to their fullest extent, however, they closed gently and without noise. Other, and probably new species of this curious genus of Crustacea I afterwards met with at Labuan; and I also took specimens at Singapore, where they were full of spawn in the early part of April. The peculiar clicking apparatus, although deserving of remark, is by no means unusual, and is shared by another genus, Alope.

A beautiful banded Comatula, or feather-star, came up with

Comatula.

the anchor, which, however, was sufficiently simple in its form to allow of its being depicted with tolerable accuracy before it performed its usual suicidal operations. These singular animals were usually of so complicated a form as to defy any attempt at drawing them with a satisfactory or useful result; and they were at the same time so perishable, that no means of preserving them could be found. If placed in fluid they discharged all their colour into it, and became so brittle as to break up into minute fragments; and if dried, they became of an uniform blackness, and were inevitably broken more or less in the process.

The weather having at length cleared, and our surveying operations being completed, we quitted Haitan Straits on the 2nd July, and steering northward, lay off the White Dog Island for the night. A curious effect presented itself as we approached the mouth of the river Min on the following day, in the sudden change of colour of the water. About two or three miles from the coast, and parallel with it, ran a long and well-defined line, stretching as far as the eye could reach, the outer side of which was distinguished by the greenish tinge of sea-water near shore, while on the landward side were the yellow turbid waters charged with the mud brought down by the river Min. The line of demarcation might have been drawn with a ruler, and a single step would have sufficed to cross it.

As we were anchored off Woga Point, in the mouth of the river, and lay there a day or two in order to procure fresh provisions, I took the opportunity of ascending the river as far as Pagoda Anchorage, about 10 miles short of Foo-chow-foo. This is a very picturesque excursion: lofty hills arise on either side nearly all the way, their slopes sweeping boldly to the water's edge, and terraced for cultivation often to the very summit. The soil, however, is naturally barren, the hills being of a dark stone, and in many places quarried, extensive bare patches being not unfrequent, conspicuous from the absence of herbage upon them, and glistening with the moisture which trickled over them. These rocks often sloped in smooth tables to the river's bank, and in one spot assumed a remarkable form, known to the Chinese as the *Mandarin's leg*. This is a block of stone projecting from the smooth hillside in *bas relief*, and in one aspect bearing a singular resemblance to a gigantic human leg and foot, placed obliquely, the toe touching the water as though kicking it. It is said that at this spot the influence of the salt water ceases. Near Pagoda Anchorage a light-coloured rock juts out on the left bank, formed entirely of soapstone. It descends sheer down into the river, the lower part being soft and disintegrated, and mottled with yellow and flesh-colour, and the general texture that of softish sandstone, but with veins of a substance more resembling the ornamental soapstone. Many ornaments, in the form of pagodas, &c., are made of this stone, and are sold in the neighbourhood at a very low price.

At Kwan-tau a great many junks were anchored, and also at Tin-tac; and we passed many large pole-boats, as they are termed, laden with spars attached to their outsides. These "sticks," such as are used for large junks, are often of very great size, and weigh sometimes no less than ten tons, the mainmast of our own ship scarcely exceeding five.

Chinese forts are numerous on the river's banks: from one spot at least four were visible, consisting of stone embrasures, with large guns, usually eight-inch, peeping through them, their muzzles painted with red-lead, and protected from the weather by a wooden frame. Some of these forts mounted 30 guns, others only about a dozen; some were very conspicuous, others masked with herbage. A small island in the stream bristled with fortifications and guns; but it was evidently a very old and tumbledown affair, more picturesque than useful.

Pagoda Anchorage receives its name from a stone pagoda built upon Losing Island, in a broad reach of the river, near which is the town of Mingi, a busy spot, inasmuch as it is the emporium which carries on business between Foo-chow and the shipping at the anchorage. The Pagoda is picturesquely situated, rising above trees and terraced houses, so that four storeys are visible, of a dark grey or brown colour and of rather heavy proportions. This perhaps arises from the fact that the top storey was wantonly destroyed some years since by some sailors of a man-of-war, so that it naturally has an imperfect and unfinished appearance. Not very far from the spot a second pagoda may be seen from the river, up one of the vistas between the hills. That these pagodas have some religious meaning I think cannot admit of a doubt, notwithstanding that a Chinaman, when I questioned him, laughingly denied that they had anything to do with "joss pidgin." But anyone who has mixed with the Chinese soon discovers two peculiarities in their character, viz., first, that they will laugh off any inquiry or reference to their religious customs or superstitious belief, as though it were not worthy of discussion, or too ridiculous to be mentioned; and secondly, that it is impossible to get any satisfactory answer to rational inquiry into their own manners, customs, politics, or history. I never heard of any reliable information being obtained from a Chinaman upon such points in a general way; and it is only those who have long lived among them, and have mastered not only their character but also their language, who can place the least confidence in the statements made by them.

From what I could gather concerning pagodas, however, they were decidedly of a religious or superstitious origin. All the country round which can be seen from the top of one of these structures is supposed to be blessed with fertility, and the higher, therefore, and the more imposing the building, the greater would be the extent of rich and fertile district; and I have been informed that the inhabitants of all the villages of a given neighbourhood united their means in order to cause a respectable pagoda to be constructed, which should thus bring a common blessing upon them all. Nine storeys, however, seem to be a favourite number, and most pagodas appear to be of greater or less antiquity. At the anchorage rode about 30 vessels of all nations, among which several Americans were conspicuously dressed out with flags, it being the 4th July.

We returned to the ship the same night, narrowly missing it in the dark owing to the extraordinarily rapid tide which flowed out, and against which pulling would have been of no avail; and the next day we left our anchorage at the mouth of the Min, and, crossing the remarkable tidal line before mentioned, made for the Incog. Islands, about which some surveying operations were attempted; but a strong northeast wind drove us off the ground.

The coast of China, though barren and desolate, is not without picturesqueness. The numerous islands which are clustered along nearly its whole extent are seldom visited by European ships, partly because they have ever been the nests of pirates, and also because the navigation would be far too intricate to be of advantage. A good idea of their character was gained by passing through the Shwin-gan Passage, in which the islands assumed a comparatively fertile aspect, being green and dotted with trees, the ground broken and hilly, with here and there a few bullocks grazing. It was like steaming along a pleasant lake, with just water enough to float the ship. One or two large villages lay at the base of the hills, the occupation of whose inhabitants was denoted by the numerous large triangular nets which lay spread out upon the slopes, and were visible, from their numbers, for a considerable distance. The narrow exit was staked for fishing purposes; but we managed to pass through a small gap without injuring them. The numerous cuttle-bones floating outside were probably indications of the same thing, for they use these animals largely as bait; and beside them were Physaliæ, Velellæ, and other marine animals.

This night was one of the most wild and weird-looking that I ever witnessed. There was no moon, and it was very dark; but the sea was highly luminous, every wave breaking with a pale light, which rendered it visible at a considerable distance, so that the whole sea was streaked with unearthly fire, and the ship was enveloped in a sort of luminous sheath. Vivid lightning flashed incessantly all around, momentarily rendering the scene more strikingly wild; and the ship rolled all the time so violently that it was almost impossible to walk the deck. The combination of effects produced a mingled feeling of awe and delight, and realised the wild dreams embodied in some passages of Dante or illustrations of Doré.

Far out to sea we came upon some small open boats, which one would have thought only fit for crossing a ferry, and in which some Chinamen were tending an extensive series of floating nets, extending from one-half to three-fourths of a mile in length. The industry and enterprise of these people were strikingly exhibited, and a feeling of surprise and admiration was excited by the extent of their operations, and the boldness with which they carried them out.

It was here, when we had stood out some 30 miles from the land, that a plague of flies overtook us. The cabin was so full of them that the rafters were blackened. Common black house-flies, for the most part, with, however, a good sprinkling of large green flies. Where they could have come from was a mystery; but they were a terrible nuisance, and although we swept off hundreds in a butterfly net, their numbers did not appear to be sensibly diminished.

Another singular circumstance was, that although no land was in sight, large dragon-flies repeatedly flew across the ship; and I observed a large dark butterfly flit across in the direction of the land, without stopping to rest on the ship. At this time the nearest land was the Chusan Islands, fully 30 miles distant. It is by no means an uncommon circumstance to see butterflies launch themselves off one shore for a short aërial excursion to the opposite shore, half a mile or a mile distant, without the least hesitation; and when anchored in such a harbour, as at Ke-lung, they were constantly flying through the rigging so rapidly that it was impossible to catch them, for they never rested upon the ship. Under these circumstances, they usually fly low, in a straight line, and near the water.

Soon after passing through the Fisherman's Group of islands, the discoloured yellow and muddy water announced that we were approaching the embouchure of one of the great rivers of the world, the Yang-tze-kiang. The entrance is anything but imposing, the coast being perfectly flat, only relieved by a few trees and sheep dotting the landscape. The delta at its mouth consists of three islands, the passages between which are closing up, so that they will ultimately become one. But we had not much to do with the great river, the broad vista of which we saw in perspective before us, between the boundary marks of its low mud-banks —but, turning to the left hand, we passed into the Wo-sung river, upon which the city of Shanghai is situated. At its entrance are the remains of a fort or battery of 130 guns— the long line of embrasures, dismantled and in ruins, having been captured by the English in the late war. Having passed the port of Wo-sung at the mouth of the river, there is absolutely nothing to see upon the low banks; and the only thing which was worthy of attention was a sight only too common in China: lying on the mud, clothed, just as he had been cast up by the remorseless stream, was a dead Chinaman, and over him were standing ravenous dogs, devouring the carcase, and fighting over the unconscious victim as they would over a bone—the larger driving away the smaller, and digging his muzzle into the ribs of the unfortunate corpse. And people were working in the fields within a stone's-throw of this disgusting sight!

Presently afterwards we were at anchor at Shanghai. The hot season had just begun, and the sun's power was terrific. And the close, narrow, crowded river seemed stifling after the free open sea to which we had so long been used. After a few days spent in this great northern city, into the characteristics of which I must not enter, I took leave of H. M. S. "Serpent," which was ordered north, and retraced my steps to Hong Kong.

The Waterspout.

CHAPTER X.

HONG KONG TO LABUAN.

Atmospheric Phenomena—Fiery Cross Reef—Corals and Coral Fish—A Wade on the Reef—Marine Animals—Gigantic Anemones—Anemone-inhabiting Fish—Stormy Weather—Waterspout—Aspect of Labuan—Vegetation.—The Jungle—Camphor Trees—The Coal Mines—Workings—Quality of the Coal—Geological Considerations—Petroleum.

AFTER a rapid passage back through some of the spots where we had so lately lingered, and which had become so familiar—and a few days' preparation in the Anglo-Chinese colony, we once more quitted Hong Kong, on the 23rd July, and this time for the south. It being the middle of the typhoon season, we made the best of our way across the region in which they might be looked for; and, after a most enjoyable cruise of nine days, reached the latitude of 10° N., where we were safe from those scourges of the

China seas. The weather was magnificent—calm and bright—and although we crossed the path of the sun on the 25th, the heat was not oppressive. Nothing remarkable occurred; but a succession of clear brilliant days, with gently undulating seas, and grand piles of cumulus in the sky, tier upon tier, down to the most distant horizon. At night, too, the light of the full moon made it almost as bright as day, and the constellations shone with a brilliancy only to be seen in these latitudes, the brighter stars like clear lamps, leaving a trail of light over the smooth sea. I will not dilate upon the beautiful atmospheric effects which succeeded each other with never-failing variety; but will only allude to one circumstance, which struck me as unusual. The sun was going down almost cloudlessly in the west, when I observed arising from a point in the eastern horizon, directly *opposite* the sun, those beautiful radiating bands of light so commonly noticed around the sun itself, and popularly known under the expression of the "sun drawing water." Their occurrence opposite the sun, however, struck all whose attention I directed to the phenomenon, as something novel and unusual.

On the 1st August we anchored at the edge of an extensive coral reef, marked on the charts as Fiery Cross Reef, from the circumstance of the ship "Fiery Cross" having been wrecked thereon. The surface of the sea was perfectly smooth and glass-like, so that at the depth of 60 or 70 feet we could see the anchor lying at the bottom among blocks of coral as distinctly as if it had been but six feet from the surface. Never to be forgotten is my first ramble over this coral reef on such an afternoon. Taking a boat, with a couple of rowers, I left the ship and steered in search of the shallowest portions of the coral-strewn sea. A short row brought us upon a two-fathom patch, over which I allowed the boat to drift slowly; and leaning over the side and looking down into the mirror-like sea I could admire at leisure the wonderful sight, undistorted as it was by the slightest ripple. Glorious masses of living coral strewed the bottom: immense globular madrepores—vast overhanging mushroom-shaped expansions, complicated ramifications of interweaving branches, mingled with smaller and more delicate species—round, finger-shaped, horn-like and umbrella-form—lay in wondrous confusion; and these painted with every shade of delicate and brilliant colouring—grass-green, deep blue, bright yellow, pure white, rich buff, and more sober brown—altogether forming a kaleidoscopic effect of form and colour unequalled by anything I had ever beheld. Here and there was a large clam shell (Chama) wedged in between masses of coral, the gaping, zigzag mouth covered with the projecting mantle of the deepest prussian blue; beds of dark purple, long-spined Echini, and the thick black bodies of sea-cucumbers (Holothuriæ) varied the aspect of the sea bottom. In and out of these coral groves, like gorgeous birds in a forest of trees, swarm the most beautifully-coloured and grotesque fishes, some of an intense blue, others bright red, others yellow, black, salmon-coloured, and every colour of the rainbow, curiously barred and banded and bearded, swarming everywhere in little shoals which usually included the same species, though every moment new species, more striking than the last, came into view.* Some, like the

---

* A very distinguished zoological writer, when speaking of a somewhat similar scene, remarks, "It is excusable to grow enthusiastic over the infinite numbers of organic beings with which the sea of the tropics, so prodigal of life, teems: yet I must confess I think those naturalists who have described, in well-known words, the submarine grottoes decked with a thousand beauties, have indulged in rather exuberant language." With these lines before me, I

little yellow chætodons, roamed about singly; others, in large shoals; some were of considerable size, and seemed to suck in the little ones like motes in the water; and in an interval a small shark, about ten feet long, swam leisurely past. A baited hook hanging over the stern attracted several species, which nibbled harmlessly at it; while many others paid no attention to it; and it is somewhat singular that although I took several, they were all of one out of the numerous species which were gliding in and out of the sheltering branches of coral. At the same time, from the ship, several large fishes, known to the sailors as *snappers*, were taken—bright red, with large scales, hard fins, and several sharp teeth; but, according to usual experience, these could only be captured in the first half hour or so after the ship had come to an anchor.

On subsequent occasions a ripple upon the surface of the water destroyed the great charm of the reef as I have thus faintly described it. With the aid of the water-glass, however (a long tube with a thick piece of glass let into the lower end), this difficulty was in a great measure obviated. The reef proved to be very extensive; but in most parts not less than two fathoms under water, and in no part awash. On one of the shallowest patches, three miles from our anchorage, were the timbers of a wrecked ship, the "Meerschaum," her iron stanchions sticking several feet out of the water, and visible as a landmark for a considerable distance.

Before leaving Fiery Cross Reef, however, I found a tract which was not more than three or four feet under water, and

cannot alter one word of what I have written. I can only say, that although I do not deny a certain amount of enthusiasm, the scene I witnessed on this reef fully justifies the language I have used in an attempt to describe it. An officer of the ship who was with me, but no naturalist, was equally warm in his expressions of surprise and admiration.

only sparsely covered with coral. Here, accordingly, having taken a boat with a couple of Malays, I jumped overboard, and, in spite of sharks, waded about breast, and even neck, high. Under the coral blocks were numerous Ricinulæ, Turbones, Turbonilli, and a few cowries and small cones, mostly encrusted with Nullipores; but nothing remarkable in the way of shells rewarded my search upon this occasion. Numerous small Tunicates adhered to the stones; and upon them I also found two species of Nudibranchiata which I had not previously met with. One of these was new, perhaps a species of Chromodoris, and which occurred on no other occasion; the other, a remarkably tuberculated animal, having but little of that beauty of form and colour which distinguishes the family, was perhaps a variety of the Doris exanthemata of Kelaart, found by him upon the coasts of Ceylon. I afterwards met with other specimens of this Doris on the coast of Borneo, where it attained a magnitude and degree of unsightliness which astonished me. It was there nearly eight inches long, of an olive-green colour, adorned with bosses and tubercles, which rendered it anything but a pleasing object. It was also extremely slimy; and Kelaart remarks that it is impossible to preserve this species in spirits, owing to its being semi-gelatinous and rapidly dissolving when dead. I did not find this to be the case, however, for all the specimens I found were readily preserved, and hardened in the spirit. This is one of the largest Nudibranchs recorded; but the specimen I found upon Fiery Cross Reef was only three inches long.

The Malays at once commenced wrenching out the great clams, which they called *chama*, avowing that they would make excellent curry; but when I directed their attention to the numerous black Holothuriæ which lay scattered about in profusion, they replied, "Chináman eat Trepang, Malayu no." I collected some, however, and found upon them numerous Stilifers. Echinoderms were, however, few. There were no starfishes; but a delicate, long-spined Echinus (Calamaris annulata), which moved very nimbly by means of attenuated suckers, so as to elude capture by creeping into crevices from which it was impossible to dislodge it without breaking the tender, barred spines. It was not easy to touch the spines of this species ever so lightly without being wounded. Even when carefully approaching them I have found the spines sticking into my hand; and I was almost persuaded that they had the power of ejecting some portion of the spine as a means of defence—a persuasion which I afterwards found amounted to a belief in the minds of some who had met with this species upon the reefs at Labuan, where it also occurs.

There were but few crustacea, and the most interesting was a new species of the elegant genus Melia, small in size, and having the carapace tesselated, and delicately painted with black, red, and yellow. Mr. Spence Bate has named it M. grossimana.

But by far the most remarkable circumstance I met with on the Fiery Cross Reef was the discovery of some Actiniæ of enormous size, and of habits no less novel than striking. I observed in a shallow spot a large and beautiful convoluted mass, of a deep blue colour, which, situated as it was in the midst of coloured corals, I at first supposed to be also a coral. Its singular appearance, however, induced me to feel it, when the peculiar tenacious touch of a sea-anemone made me rapidly withdraw my hand, to which adhered some shreds of its blue tentacles. I then perceived that it was an immense Actinia, which, when expanded, measured fully two feet in diameter. The tentacles were small, simple, and very numerous, of a deep blue colour; and the margin of the tentacular ridge was broad and rounded, and folded in thick convolutions, which concealed the entrance to the digestive cavity.

While standing in the water, breast high, admiring this splendid zoophyte, I noticed a very pretty little fish which hovered in the water close by, and nearly over the anemone. This fish was six inches long, the head bright orange, and the body vertically banded with broad rings of opaque white and orange alternately, three bands of each. As the fish remained stationary, and did not appear to be alarmed at my movements, I made several attempts to catch it; but it always eluded my efforts—not darting away, however, as might be expected, but always returning presently to the same spot. Wandering about in search of shells and animals, I visited from time to time the place where the anemone was fixed, and each time, in spite of all my disturbance of it, I found the little fish there also. This singular persistence of the fish to the same spot, and to the close vicinity of the great anemone, aroused in me strong suspicions of the existence of some connection between them.

These suspicions were subsequently verified; for on the reefs of Pulo Pappan, near the island of Labuan, in company with Mr. Low, we met with more than one specimen of this gigantic sea-anemone, and the fish, so unmistakeable in its appearance when once seen, again in its neighbourhood. Raking about with a stick in the body of the anemone, no less than six fishes of the same species, and of various sizes, were by degrees dislodged from the cavity of the zoophyte, not swimming away and escaping immediately, but easily secured on their exit by means of a small hand-net. Thus the connection existing between the fish and the anemone was demonstrated, though what is the nature and object of that connection yet remains to be proved.

There are at least two species of these anemone-inhabiting fish; and a second species of the same genus differs from that just described in having black and cream-coloured vertical bands, instead of orange and white. Such a fish I have seen, evidently related to the first-mentioned, living in a tub which did duty for an aquarium, in the possession of Mr. Low, at Labuan, and which had been obtained from what was probably a second species of fish-sheltering anemone. This fish was remarkably lively and amusing, and of a disposition I can only describe as *knowing;* and lived in good health in this tub for several months—a proof that the connection between these animals, whatever its nature, is not absolutely essential for the fish at least.

But the fine weather which permitted such operations on a submerged coral reef in the centre of the China sea was not destined to endure much longer, and a strong southwest monsoon now warned us away from this interesting spot. For the next week we experienced a series of squalls, accompanied by heavy rain and otherwise dirty weather; which well illustrated the difficulty and vicissitudes of the naturalist's work on board ship. The living things I had brought from the reef to draw and examine, more particularly the more delicate species, were capsized from time to time by a heavy lurch of the vessel, upsetting the salt water among papers and drawings, as well as into the colour-box, which the contracted limits of the cabin often necessarily brought into dangerous proximity. The rolling motion made it necessary to suspend operations, and meanwhile the animals died, or were lost. To write, even, was almost impossible; and to draw or use the microscope quite so, even if the lively movement of the ship had left the observer in such a state of bodily comfort as to enable him to continue work. But the unstable equilibrium of every article in the cabin kept the mind constantly in a state of alarm, and the necessary operation of *wearing ship* from time to time was fatal to minute animals and everything that could not be securely fastened; and when this operation occurred in the middle of the night, and I heard my glasses, saucers, microscope, &c., rushing from side to side of the cabin, in total darkness—every one too busy to attend to my cries for a lantern—a deluge of rain pouring down outside, my condition may be readily imagined. I gathered up the *débris* of Fiery Cross animals; but for a week all serious work, even to writing up my diary, had to be suspended; and every one who has been in similar circumstances knows how difficult it is to recover time lost in this manner.

For three days and nights we ran through the dangers of the China sea without sights, and were not sorry at the end of that time to get a glimpse of the sun and stars, and once more to verify our position. On the 7th August a succession of rapid orders upon deck made me look out of my cabin windows, when the cause was visible in the form of a waterspout, little more than a cable's length from our quarter. It had just formed at a distance of not more than two ships' lengths astern, and had slowly crossed the vessel's wake. The long, black, flexible pipe was clearly defined upon the murky background, slightly undulating—now straight, now somewhat serpentine—the broad, funnel-shaped top descending from a dense cloud, and the terminating point partially concealed in a whirlpool below. From the upper part, and from the edges of the spout, could be seen the water streaming down in torrents; while the sea below was lashed into foam, and a spiral eddy of turbulent and foaming water, rising above the level of the sea in the form of an inverted basket, received the point of the spout. On the outside of this vortex the waters could be distinctly seen whirling madly round from left to right, with great rapidity; and the whole phenomenon—cloud, spout, vortex, and all—moved majestically onward, and, having lasted about five minutes, gradually faded away—a grave and impressive sight, which will not soon be erased from my memory.*

On the 10th August we approached the low, jungle-covered shores of the island of Labuan, by no means prepossessing in appearance. The harbour of Victoria, in the south-east corner of the island, is the entrance of a river running a short course into the interior; and at the anchorage is situated the Bazaar, a collection of native shops—Malay, Kling, and Chinese, but chiefly the two latter—and also the Government offices, wooden shed-like buildings, but little ornamental, although sufficiently commodious in their internal arrangements. These, and all the houses of the European residents, are built upon piles raised five or six feet above the soil, which is damp and malarious. They are scattered at intervals over this south-east portion, which is free from jungle, and in many parts planted with cocoa-nuts, betels, and other useful trees; while the parts not so cultivated are swampy, and covered with low bushes of Melastoma, and fern of a species of Pteris, much resembling our common Bracken.

* The engraving at the head of this chapter is from a careful drawing I made immediately afterwards, from a sketch and memoranda jotted down at the time.

Where the ground is most moist two species of Nepenthes (pitcher plants) occur in abundance—one having a large mottled cup and lid, and the other a slender pitcher, of a green colour. In all the drier parts a species of grass abounds, which becomes a perfect nuisance to the pedestrian, from the fact that its barbed seeds detach themselves at the slightest touch and stick into the clothes, particularly if they be of woollen or flannel, in such profusion as to be very irritating to the skin. After a walk the feet and legs bristle like a hedgehog from the innumerable little spears, which, from some textures, can be best removed by scraping with the back of a knife; while from other materials they require to be laboriously extracted one by one, at a great exercise of patience. The English ladies call it *love grass*, from its sticking qualities.

No portion of the island is very elevated; and the country has a desolate appearance, the soil being sandy and very loose, and the vegetation heath-like. In other parts, where the jungle has been cleared, a thick tangled undergrowth flourishes; but the few trees left here and there, having lost the shelter of the jungle, have died, and rear their naked and giant arms into the air, only adding to the dreary effect of the scene. In this part of the island the most agreeable features are the cocoa-nut plantations, and the Casuarinas which grow by the roadside; but the real untouched portions of the jungle have a peculiar charm of their own. The trees in these situations attain truly magnificent proportions; and no more splendid sight of the kind ever met my eyes than in the midst of the clearance at the coal workings on the north end of the island. Standing upon a slight eminence, whichever way we turned our eyes they were met by an almost impenetrable wall of lofty and well-proportioned trees, on the verge of which the axes of the woodcutters reverberated most musically in the dark forest depths; the loud chirp of the cicadas, and the prolonged hum or whistle of numerous beetles and other insects, formed a pleasing accompaniment. Giant camphor trees (Dryobalanops camphora), Dammar, and other trees, rose straight and erect to the height of from 150 to 200 feet, clothed with foliage, or standing dead and dry in the cleared spots, or sometimes stretched at length upon the ground in wild confusion. The mere stumps were objects of wonder for their massiveness, with enormous wings or buttresses, which required long days of hewing to separate them from their iron-hard roots, which could be traced tortuously winding through the soil for sixty yards from the tree, and even at that distance were as thick as a man's thigh. Hard, close-grained, solid timber lay there in profusion, piled trunk over trunk, the greater part requiring to be slowly and laboriously burned before it could be finally removed; while the tall standing trunks all round towered up sombre and solemn, awaiting the doom of the axe. Nowhere have I seen such glorious jungle as in this part of the island of Labuan.

Many trees of the jungle belong to the natural family of the Dipterocarpeæ, a family remarkable for beautiful flowers no less than for their majestic size, erect trunks, and fine dense foliage; and not a few yield some kind of balsamic resin. The form it assumes in the Dryobalanops, or Sumatra camphor tree, is that of concretions in the crevices and fissures of the wood, so that it can only be obtained by cutting down the tree, which, inasmuch as they are often 90 feet high, without a branch, is no small labour. When felled, the trunk is hewn in pieces, and the camphor found in clear crystalline masses, and with it an essential oil known as camphor oil, which is believed by some to be camphor in an imperfectly formed condition. Although, however, the oil is artificially crystallised, it does not produce camphor of so good a quality as that which is found already solidified in the cavities of the wood. The camphor which has already been alluded to as obtained from Formosa, has a different source, viz. the Laurus camphora, or Camphora officinarum, a tree of the laurel family, in which wood, branches, and leaves, alike yield camphor by dry distillation—as it were, a solid evaporable oil. This commercial camphor is, however, more volatile than the hard camphor obtained from the Dryobalanops, and its presence as a vegetable secretion is not confined to the true camphor laurel, it being also found in other lauraceous plants, especially cinnamon.

The Dammar trees are Coniferæ, of the genus Dammara, yielding a hard and brittle resin, like copal, which is itself a Dipterocarpous product. A substance similar to Dammar, however, appears to be produced by several kinds of trees in these forests.

It is in this dense jungle that the mineral wealth of Labuan so long lay concealed; but the coal crops out so conspicuously, less than half a mile from the sea-shore, that it is no matter of surprise that it attracted notice. The district containing the coal-beds is composed for the most part of a soft yellow sandstone, which dips 33° N. by E., and the coal exists in several seams, of which the largest is no less than 11 ft. 4 in. in thickness, though the quality of this seam is by no means so good as that of some thinner ones. The coal roof is a stiff blue clay (not a fire-clay), and beds of shale alternate with the seams. The uppermost seam, called No. 1, is 4 ft. 6 in. in thickness; the next is 2 ft. 9 in.; the third, 3 ft. 9 in.; and the fourth is the thick seam of 11 ft. 4 in. Between the third and fourth seams are 8 fathoms of grey shale, in which fossil shells are occasionally found. I had great difficulty, however, in procuring any fossils. I could not learn that any had been kept while the excavations were going on; and only succeeded in obtaining two bivalves from the 8-fathom bed of grey shale.

The Labuan coal mines are worked in a way which offers a striking contrast to the mode of working the Ke-lung mine in Formosa. There are here two shafts at present constructed; one of these (called the Shallow Pit) enters the uppermost seam, while the other penetrates to 45 fathoms depth; a third also is in course of construction, which will reach to the depth of 100 fathoms. Besides this, there are seven or eight level workings. The great difficulty here is to obtain labour sufficient to develop the resources of the mines; for although 600 men were on the books, including Chinese, Malays, Klings, &c., with European departmental superintendents, only 300 were at work at a time. At the time of my visit 80 tons per diem were produced, and conveyed down a tramway which descends to the coaling pier; but, with more labour, I was assured by the manager that the same machinery would produce 200 tons per diem with ease; although that, of course, is not a large quantity compared with the diurnal supply from mines at home.

The quality of the Labuan coal is superior to that of Ke-lung. It is a heavier, closer-grained, though not very clean coal, is very free from sulphur, and forms but little clinker, in this respect having a conspicuous advantage. It burns, however, very fast, and generates steam very readily as long as the tubes and the fires are clear; but the former require cleaning every 24 hours, and the latter every eight hours. Owing to its rapid combustion, it gives out a very considerable amount of heat, so that it is necessary to be careful that the red-hot flues are protected and watched, while the flames issuing from the funnel extend sometimes six or eight feet, and endanger the rigging. In burning it produces a large quantity of soot and imperfectly consumed fragments, which render everything gritty and dirty. Still it is better than Ke-lung coal; and I have heard several engineers assert that a mixture of half Labuan and half Welsh coal forms a fuel which is excellent for all practical purposes. But being a much inferior coal to Welsh, it naturally requires more room for stowage, the difference being well expressed by the fact that Her Majesty's ship "Scylla" consumed in 24 hours $37\frac{1}{2}$ tons of Labuan coal instead of 27 tons of Welsh; and again, the average quantity of water distilled in the condensers by one ton of Welsh coal was 7·2 tons, while Labuan coal, although mixed with one-third of Welsh, only condensed 6·25 tons. There are various qualities of Labuan coal, however; the best is far inferior to Welsh; and though the quantity of ash is small compared with that left by Ke-lung coal, the amount of soot and unconsumed carbon is very large.

Labuan coal is supplied by the Company to ships of war by contract with Her Majesty's Government, at the rate of 1*l*. per ton, and to merchant vessels for 1*l*. 5*s*. The Labuan Coal Company have depots in Shanghai, Hong Kong (where it fetches 2*l*. 2*s*. per ton), Singapore, &c.; and they are anxious to extend their operations.* But it must be con-

---

* At the time I was at Labuan the prospects of the China Steam-ship and Labuan Coal Company looked rather dark, and they have, since the above was

fessed that merchant ships prefer to purchase British to Labuan coal, although the former is much dearer.

The physical characters and geological relations of Labuan coal seem to point out that it, like the Ke-lung coal, is a recent formation; in fact, a lignite. In the stiff clay roof of certain seams, Mr. Low assures me he has found many impressions of leaves in very perfect preservation, identical with those of trees growing in the jungle at the present day. In the coal there are very frequently found tears of pure dammar resin, and the Dammar trees are still common in the jungle. This resin has also a great tendency to occur in veins; and I was informed that on one occasion a mass of pure dammar 6 lbs. in weight was found. It was intended that this unique specimen should be deposited in an English museum, but a careless workman unfortunately let it fall, and it was broken to fragments. The Dammar trees will exude resin while in a state of decay, and so long as a particle remains undecayed, that particle will continue to produce its quantum of resin. Mr. Low informed me that he knew of trees which for 12 years have been undergoing a slow process of decay, and that the resin is yet distilled from the small particles which remain.

Two or three miles south-west of the coal mines, in a deep *nullah* of the jungle, a petroleum spring has been discovered, and a path has been cleared through the forest to the spot, which I visited. This path runs along a high ridge, from which a grand view is obtained of the majestic trees whose lofty stems ascend from the deep valley on either side. Among the branches of these trees, monkeys

written, wound up their affairs. I understand, however, that a new Company has been formed, and that the coal continues to be worked; and there seems to be no reason to doubt that with judicious management, the work will be carried on advantageously and successfully.

sported, birds sang, and shrill cicadas made the forest re-echo; beautiful butterflies flitted across the open glades; scarlet-bodied dragon-flies shot hither and thither like painted arrows; and now and then curious green mantises, and walking-stick insects (Phasma), scarcely distinguishable from the green and brown twigs, arrested my attention.* No workings have yet, however, been undertaken at the petroleum spring; but as I have learned that others exist in the neighbourhood, no doubt they will one day form important sources of material and revenue.

* I was rewarded in this walk by the capture of a handsome new species of Phasma.

CHAPTER XI.

LABUAN.

Bruni, the Capital of Borneo—Piracy—Establishment of the Colony of Labuan—Its Objects—Natural Productions—Pigs—Monkeys—Kahau, or Proboscis Monkey—Birds—Megapode—Chick-chack—Barking Lizard—Iguanas—Cobra—Pythons—Electric Snake—Scorpions—Centipedes—Cicadas—Beetles—Hemiptera—Desecration of European Graves—Isolated Position of the Residents of Labuan.

THE island of Labuan was ceded to the English Government in 1846 by the Sultan of Bruni, the nominal ruler of Borneo. Bruni, the capital of Borneo, and from which the whole island takes its European name, situated a short distance south of Labuan, is a town of some 25,000 inhabitants, governed, or rather misgoverned, by a Rajah and subordinate chiefs, whose sole aim is their own aggrandisement, and the increase, by fair or foul means, of their own revenues. By a system of peculation and persecution, which is for the most part delegated from the chiefs to certain inferior dependents, the ill effects of their rapacity are spread through the population in such a way that few escape, and justice is a commodity almost unknown, unless it happens to side with self-interest. Were it not that the nobles themselves find it impossible to agree, and therefore do not unite largely in their avaricious projects, the country would almost be reduced to a state of savagery and unproductiveness; as it is, it is only saved by a kind of patriarchal feeling, which keeps each section of the city together, and unites family connections especially, by a strong *esprit de corps*.

The terrible prevalence of piracy upon these coasts attracted the attention of philanthropists no less than of merchants, from the destruction of life and property, and the insecurity of ships trading in those seas; and although the Sultan of Borneo had no doubt a direct interest in the success of piratical operations, he was forced by cogent arguments to listen to reason, and to enter into a treaty with the British Government, the basis of which was the suppression of this scourge. In 1849 a similar treaty was concluded with the Sultan of Sooloo; and these treaties, backed by some severe practical lessons, in which fleets of piratical *prahus* were dispersed or destroyed, have by degrees very considerably mitigated this terrible and lawless trade, for which benefit the world is in the first instance indebted to Sir James Brooke, Rajah of Sarawak.

It seemed desirable that a British colony should be established somewhere upon these coasts, which might be used as a naval station such as might hold some check upon these lawless proceedings, and keep a wholesome fear before the minds of an island population in whose eyes piracy and murder were profitable trades, and no crime; while at the same time an advantageous result would ensue in the opening up and development of the commerce of a vast country whose resources were very imperfectly known, but must of necessity, from its magnitude and geographical position, be enormous. Added to this, the existence of coal in unknown quantities upon the island of Labuan pointed out that spot as most suitable for the proposed settlement, and a treaty was accordingly made, by virtue of which Labuan became at once an appanage of the British crown, a harbour of refuge in the China sea, a basis of operations for the suppression of piracy, a centre of development for British commerce with Borneo, and a great and almost boundless source of fuel for the supply of ships trading in the East.

These great ends, which appeared to be promised by the colonisation of Labuan, have not perhaps all been accomplished; nor has the settlement flourished to that extent which this enumeration of its advantages would appear to have warranted. Trade has not increased with Borneo at the rate which sanguine persons hoped; and the coal resources are perhaps scarcely of that quality which it was at one time hoped they would prove to be; while the climate of the island is insalubrious, and ill-suited to Europeans. The proof that the island is not attractive either to commerce or to adventurers is to be found in the fact that there are no European residents, except the Government officers and those directly employed in the working of the coal mines, and these with one accord would be glad to leave it. And if report indeed speaks truly, the advantages of the colony are so little preponderant, that it is a question in the minds of those in authority whether or no the colonial establishment should be abandoned.

But however these questions may agitate the minds of legislators, of one thing I became satisfied, viz., that, for a short residence, the island of Labuan was a place of extreme interest to the naturalist; and as I had the opportunity of spending some weeks here ashore I shall devote some space to an account of its leading natural-history features as they presented themselves under my observation. Although very low and flat, as a rule, the island is by no means devoid of diversity, owing to the prevalence of jungle and the proximity of most points to the sea; and indeed there are pleasant undulations in certain parts, as in the neighbourhood of Government House, and at the north part of the island, which redeem the general tameness of its surface. One circumstance, which is not without its advantages, is, that there are no wild animals which are liable to pounce upon the unwary wanderer in the woods, armed perhaps with no more deadly weapon than a butterfly-net. The largest quadruped here is a pig, wild and black, but harmless, and keeping as much as may be out of sight. I rarely indeed met with them, although they no doubt abound, and on the mainland attain a very considerable size, standing from 3 feet to 3 feet 6 inches high, and, when wounded, being very savage and highly dangerous.

The jungle of Labuan abounds also in monkeys, which are all, however, of small size, and ornamented with long tails. A walk can scarcely be taken in the woods without meeting some of these animals leaping about in the trees, and chattering at the intruder. Not unfrequently they boldly leave the jungle and approach the houses, probably for the sake of finding something edible in the gardens; and the first I saw was from a verandah, looking across a plot of grass which overhung the sea-beach. The Klings pay them religious honours; and one belonging to an officer of the garrison, having escaped, and defied every effort at capture, it became necessary to shoot it, as it amused itself by destroying the *attap* roof of the house. The Sepoys begged the body, and having received it, they dressed it up, and having paid it burial honours, concluded the farce by burning the carcass upon a funeral pile.

None of the larger, or tailless species, occur upon the island, although upon the adjacent mainland some species are not uncommon. The red orang or *Mias* (Simia satyrus) is the most remarkable of these; but appears to be local in its distribution, and in a manner difficult to account for. I was somewhat disappointed at seeing no sign of them up the Sarawak river; and afterwards learned that they never frequent that river, though they descend the neighbouring Sadong river in the fruit season; and I met with those who had frequently seen them there, and who assured me that they measured fully five feet in height. Fortunately, however, they do not attack man without provocation; though, if molested, they will show fight, and hand to hand encounters not unfrequently take place between them and the owners of the fields which they invade. Their teeth are then their chief weapons of defence, and it is no uncommon thing to see a man who has lost a finger or two in these unequal combats.

Within two miles of Labuan, and between it and the mainland, is a small island, Pulo Daat, which, on the formation of the settlement was covered with jungle, in which that very handsome and remarkable species the Kahau, or Proboscis Monkey (Nasalis larvatus) at one time abounded. It is called *Bangkátan* by the Malays, and is well known upon the adjacent coast. This beautiful species has never been seen alive in England, and the specimens of it which are found in our museums give but a faint idea of the richly-coloured and glossy coat of which the living animal boasts. They are very shy and difficult of approach, and when disturbed are very fierce, being armed with remarkably large canine teeth. Several attempts have been made by the officials resident at Labuan to keep the Kahau in confinement. It has not been difficult to procure young animals, and unless taken very young, they never live more than two or three days in captivity. At the present time, although the island of Daat has been cleared of by far the greater portion of its jungle for the purpose of transforming it into a cocoanut plantation, and only a small corner of it is left in a state of nature, there can be little doubt that the Kahau still haunts it. I was unable to get a sight of it, owing to the dense and impenetrable nature of the thicket; but when watching, gun in hand, in the midst of the jungle, I was not unwilling to attribute a loud moaning sound, which was frequently repeated, and evidently by some large animal, to that of which I was in quest.

The birds of Labuan are numerous and interesting, though for the most part small; but as I was unable to pay special attention to them I will merely allude to them. The largest are an osprey, a large crow or raven, and the hornbills. Besides these there are numerous beautiful pigeons, and a variety of little Cinnyridæ (sun-birds), flitting about like so many butterflies. In this month (August), when the woods of England are perfectly silent, several birds were singing in Labuan. The commonest of these was a pied bird, as large as a starling, which sang cheerfully in open places. Another bird, only heard in the jungle, had a singular note with intervals wider than usually occur in bird-music; so that it was long before I could persuade myself that this sound was not produced by a man whistling in the jungle. At length I heard two or three answering one another, and was satisfied that I was listening to a bird; but I could never catch a sight of it in the shades of the forest. Another could be  often heard in the depth of the jungle uttering a loud and deep note, like boo, boo, boo, frequently repeated; but I never could get a view of the bird, or hear it otherwise than at a distance.

The egg-heaps of that curious bird, the megapode (Megapodius Cumingii), are not unfrequently found in the Labuan jungle. These birds build mounds, in which they deposit their eggs, several birds often uniting together to form a joint nursery, in which as many as 50 or 60 eggs are accumulated together and left to be hatched by the sun. This species of megapode is somewhat less than a guinea-fowl; but its eggs are fully as large as those of a turkey—long, and pointed at both ends, and of a brownish-buff colour. The birds themselves are more fully developed on leaving the shell than falls to the lot of most of the feathered tribes, running freely about immediately upon their large, strong feet, and capable of using their wings within a few hours after birth. Eleven of their eggs were brought in by a Malay, who had stumbled upon one of their mounds; of them six were perfectly fresh, while the remaining five were far advanced in incubation. Having placed them in a box for safety, they were left till the following day, when I found one hatched and fully feathered; but wishing to preserve the remaining eggs, I pierced them and left them to the mercy of the ants which freely roamed in and out the shell, but within 24 hours a second little megapode appeared, prematurely hatched, smaller, weaker, and less feathered than the first.

The commonest of the lizards of this island, and indeed of the whole region, is the little animal called Chick-chack (Ptyodactylus gecko,) so named from the chirping noise it makes from time to time, and which might at first be mistaken for the voice of a bird. They are perfectly harmless, and often very familiar. They live in considerable numbers within doors, concealing themselves upon the roofs, and among the *attaps*, or palm-coverings, or crawling about upon the walls and ceilings. I have counted as many as two dozen overhead while I have been at dinner in a good-sized room, some as long as my hand, and usually pale-coloured. They vary, however, somewhat in colour, according to food and locality. I have been informed by credible friends of instances in which they would habitually come down upon the table and take food offered to them, and it is equally certain that they occasionally come down involuntarily, losing their precarious footing overhead while in chase of an insect, in which case they fall with a thump upon the floor or table,* an accident which usually results in the loss of their tails, which break off with the shock or the fright; and it is by no means unusual to see them with their short stumpy caudal appendages in process of reproduction. Such an occurrence happening in the night I have found rather startling. If a moth or a butterfly flutters about near the ceiling, the chick-chacks are all upon the alert, running at it as it passes near them; and although the reptile may succeed in catching it, the insect is often too unwieldy for them, and they have considerable difficulty in securing it. They clear the house of mosquitoes and flies, however, and are never molested, but, on the other hand, always encouraged. A singular circumstance occurred to the colonial surgeon, who related it to me: he was lying awake in bed when a chick-chack fell from the ceiling upon the top of his mosquito-curtain; at the moment of touching it the lizard became brilliantly luminous, illuminating the objects in the neighbourhood, much to the astonishment of the doctor, who had never before witnessed such an occurrence.

Another lizard of a larger size than the last is the barking lizard (probably Gecko verus), which lives in trees and also

* A pretty little white-spotted lizard (Hemidactylus triedrus) fell on one occasion upon my knees in this manner, while sitting in the verandah.

about houses, from time to time betraying its whereabouts by a sound resembling a short growl, followed by a short sharp bark, not unlike that of a puppy at play. When I first heard it in a tree I looked up for a monkey; but on many subsequent occasions it appeared very much like the barking of a small dog. They are very difficult to detect, however, cunningly concealing themselves; and although I have watched for one which was barking a few yards above my head in a tree, I have looked in vain for a considerable time. They are very fond of coming into houses, and are considered by the Malays as reptiles of good omen. They feed upon insects and moths, being particularly partial to Sphyngidæ.

The Iguana (Varanus Dumerilii ?) is a larger species, which reaches a length of seven feet. Although in other respects harmless, it does considerable mischief among domestic fowls, frequenting the neighbourhood of houses for the purpose of robbing the hen-roosts. For that reason they are destroyed, and, moreover, by some they are considered excellent eating. While drying some marine animals in the sun one morning, an iguana appeared upon the scene, walking on tiptoe across the grass, and lifting its head as if scenting something. On my driving it away it returned again three times to the spot, although the bait was not what I should have imagined to be very inviting. I was unwilling to shoot it; but two or three days after, I fear it met with such a fate in the neighbouring grounds, where it was found near the hen-roost. It was about four feet long. I believe these animals, however, do sometimes subsist upon marine animals, which they pick up on the beach. I have more than once observed them skulking among the roots of the trees close to the margin of the shore; and on one occasion, I passed and repassed the spot several times on purpose to observe it. Each time I passed it retreated into the jungle, but was always at its post when I came back.

On one occasion I disturbed a large Iguana in such a situation, that in order to escape, it had to run some distance across an open space in my full view; and it did run tolerably quickly, but in a most ludicrous manner: the short and peculiarly situated fore legs had an awkward waddling motion, while the hind legs, seeming less encumbered, ran more quickly, and threatened to overtake the head, while the long tail followed behind, as if it scarcely belonged to it, swaying from one side to the other according to the direction the animal took. With all this, however, I should scarcely have caught it in a flat race.

Chameleons of more than one species exist in Labuan. The natives have a great prejudice against them, and will not touch them, believing them to be deadly poisonous.

Although as an island, Labuan is undoubtedly more free from serpents than the mainland, they are quite sufficient in numbers, and occasionally venomous. The Cobra (Naja tripudians) appears to be almost unknown in Labuan, only one instance having occurred in which it was supposed to have been met with. A gentleman (the Colonial Surveyor) riding on horseback near Tanjong Tarras, was confronted by a snake in the midst of the road, which raised itself, head erect, and hood swelled out. Although he had never seen a cobra, the snake immediately recalled to his mind the figures and descriptions of them. He dismounted and killed the reptile, and being then pressed for time, he threw it among the grass at the roadside, intending to return and examine it at his leisure; but for some reason or other, unfortunately neglected to do so. A reference afterwards to figures of the cobra only served to convince him that he had seen one of these reptiles; but it was unfortunate that it was never examined, and Mr. Low, who looked for it on hearing of the circumstance, could not discover it. The snake was five or six feet long, and, if a cobra, probably came over from the mainland in wood, or some cargo.

In Sarawak, however, the cobra is not uncommon, and grows to a large size. The Bishop of Labuan informed me that he had killed black cobras in his own house 9 ft. 6 in. and 10 ft. 5 in. in length, respectively. The former of these attacked a servant in one of the ground-floor rooms of the house. The man was paralysed with terror, and unable to defend himself; but being alarmed by his inarticulate cries, the bishop entered and found the reptile erect, his head broad and depressed, as large as the palm of the hand, and, as he expressed it to me, "barking like a dog." He fortunately succeeded in despatching it with a rattan.

Large Pythons (Python reticulatus) exist in Labuan, usually making their lair in a deep *nullah*, sometimes near a house, whence they probably derive the advantage of stray fowls or other domestic animals, which serve as their food. Mr. Low assured me he had seen one killed measuring 26 ft.; and I heard, on good authority, of one of 29 ft. having been killed there. In Borneo they were said to attain 40 ft., but for this I cannot vouch.

But although there are numerous species of snakes in the island, they are not often seen. I myself, during the space of a month, only met with four or five species, the most common of which were Dendrophis caudolineatus and Tropidonotus stolatus. In one of my walks through the jungle I was fortunate enough to secure a specimen of the rare (venomous) Elaps (Callophis) intestinalis.*

* A circumstance was told me by the Colonial Chaplain, Rev. J. Moreton,

On one occasion, during my visit, the Governor was sitting in his verandah conversing with the new Commandant, and was assuring him that snakes were rarely met with, and not to be feared, when they were disturbed by a noise close by. On going out to see what had occasioned it, they learned that the servants had just killed a python 12 ft. long, in the verandah—a singular comment upon their conversation.

Scorpions and centipedes are creatures which always excite the terror of those to whom they are unknown, and they do not gain any good-will by a nearer acquaintance. Persons who have not visited the tropics often imagine that one is never safe from the fangs of these venomous creatures; but, although they are not unfrequently met with, a sting is of rare occurrence,—and although painful, and followed by acute symptoms, is perhaps never fatal. There are two species of scorpion chiefly found about houses, one of a reddish colour, the more common and active of the two; and the other, a large black species (probably Scorpio costimanus), which appears to be only driven in by the weather. The latter is sluggish in its habits, and is freely and fearlessly handled by the natives, who even collect them, and place them round their necks and in their turbans, somehow escaping the penalty of their stings. Indeed these scorpions

---

which although it may seem apocryphal, I am unwilling to pass over altogether in silence. He found on one occasion outside his verandah, a snake about 5 feet long, of a reddish colour, but not mottled like a boa. It had had its head crushed, that being the usual way in which the natives destroy snakes, though it is not always immediately fatal to them, for they will crawl away after such an injury. Mr. Moreton told me that he took the snake in question by the tail with his thumb and finger, and instantly felt a strong electric shock, which ran up his arms to both his shoulders, so that he dropped the snake in alarm. Although much surprised at the circumstance, not being a naturalist, he neglected to take any means to preserve the reptile.

appear to listen to the voice of the charmer as readily as certain serpents are said to do. On the occasion of making a new road in Labuan, black scorpions were frequently turned up from under the roots of trees, and a man was always at hand who professed to charm them, and to act with them as above described. About houses, damp places —such as bath-rooms, and out-buildings—are particularly liable to be haunted by both scorpions and centipedes. But their more natural habitat is under stones, in sand, or among the roots of trees, among decaying wood, &c., though they are not unfrequently introduced into the interior of the house by the carelessness of servants, who may bring in clothes which have been laid outside to dry or to air, without having previously well shaken them.

Of centipedes I have met with several kinds, and in all the above situations. The small yellow species, similar to that found in England, is not uncommon in houses and gardens. It appears to bite, but not venomously, and with no worse effects than a nip from the forceps of an ant. One small species is luminous, like the Geophilus electricus of this country. The common and dreaded species (Scolopendra morsitans) varies very much in size; the largest I have met with being about nine inches long, and as broad as the middle finger. They have sometimes a green, and sometimes a brown colour—the former being most feared. Both these and scorpions not unfrequently occur on board ship; but it is a general belief that life (probably diet) aboard ship neutralises their venomous qualities, and that a bite from one in that situation is far less severe in its effects than from one of the same size upon land. When stung in the dark, although the offender may escape unseen, it is easy to discover to which animal the sting may be attributed

—a single puncture betraying the point of a scorpion's sting, and a double one the formidable pair of nippers on the head of the centipede. The puncture is described by its victims as similar to what might be produced by contact with a red-hot iron, and the constitutional effects are often very severe for some hours, consisting of considerable swelling, tenderness, throbbing pain, anxiety, and febrile symptoms. I have known the symptoms, after having lasted 36 hours, recur after three or four days' intermission. Ammonia is the best application.

Another species of centipede, of a large size, grey colour, with numerous legs arranged in twos on each side, I met with in the jungle on the east coast of Formosa, and also on the banks of the Sarawak river, associated with large Millepedes (Julus). It had, however, no venomous fangs.

Certainly the most remarkable insects for noise are the Cicadas. There are several species in Labuan and Pulo Daat, which make the woods resound. One of the most extraordinary of these singing insects utters a sound by no means unmusical. Just as the sun goes down, a loud, ringing whistle strikes up among the fern, or in some spot near the house, sometimes apparently almost in the verandah, which I can best compare to one smartly rubbing on very sounding musical glass, and keeping up for a long time a very loud and uninterrupted musical note. You may search in vain for the origin of the ringing sound, though it appears to spring from the very spot on which you may be standing, for a quiet approach will not disturb the insect, which, sitting in the mouth of its hole in the ground, whistles its monotonous and loud song, which is probably intensified by reverberation in the cavity. This insect seems to affect the neighbourhood of houses, and can only be seen by a patient and, withal, fortunate watcher.

But there are two or three species of Cicada which are no whit inferior in noisy powers to the insect just mentioned (which I have been assured was a locust), though their notes have a different character. One of these makes a simple chirp, chirp, all night long, like our crickets. But there are two others which I will designate respectively the *scissor-grinder* and the *saw-whetter*. I shall never forget the first time of hearing the scissor-grinder in the jungle at Pappan when approaching the island in a boat, the noise being distinctly audible for at least a quarter of an hour before we reached the shore, and when there the resounding whir-r-r—whir-r-r—whir-r-r of the insect awakening the echoes of the forest was truly astonishing. After continuing this deafening sound for some time, it winds up with a protracted whiz-z-z-z, which dies away just like the scissor-grinder's wheel when the treddle stops. Another which I heard at Coal-point closely resembled the whetting of a saw, but was not so common as the last; and a third always began with a sort of warbling note, like a person blowing in water with a bird-whistle, very loud and somewhat melodious withal. These sing all day, even during the hottest hours.

The Cicadas are, however, very difficult to detect by the sight. They often sing high up in the trees, and I should still be doubtful of the real nature of the songsters had I not once or twice, when peering curiously up into the tree, seen a Cicada quit its retreat and fly from among the leaves simultaneously with the discontinuance of the sound. But when in a bush near at hand, the ringing sound is of a peculiarly deceptive and ventriloquous nature. The noise they make is so loud that it thrills through the ears in a manner perfectly deafening. You approach the bush from which it appears to issue, and you even appear to have reached the very spot in which the animal is concealed, but, nothing daunted, the insect continues its screeching, and you may peer about and look for a glimpse of it in vain. Your proximity does not disturb it, for it seems to think that it is quite safe in its concealment, and even thrusting a stick into the bush will not dislodge it, nor in all cases even stop the noise. At the same time one cannot be absolutely certain that it is really in that particular bush, for the mere intensity of the sound is not sufficient to fix its exact locality, though the thrill it sends through the ears proves it must be very near.

The various species of beetles are nearly as successful in concealing themselves as the Cicadas, and while they are by no means exceedingly numerous in Labuan, the commonest are not in very great profusion, if we except a species of Cicindela (C. aurulenta), which flies over sandy spots, as is the habit of all Cicindelas, and could always be captured in any quantities in such situations, and the orange-spotted Dacne 4-maculata with its allies. A gentleman who had been an insect-collector for a dozen years, assured me that he had never succeeded in discovering where beetles harbour, or how to collect them in quantities. The fact probably is, that under the bark and in the decaying wood of recently-felled trees, are the situations in which we should look for them with most success. Of large species the very variable Xylotrupes Gideon is not uncommon, and I captured numerous small species which have great interest for the coleopterist.

Hemipterous insects (the bugs of the entomologist)

occur more gregariously, and were often found in abundance upon particular plants. Upon a spreading, bushy Labiate could always be taken Cyclopelta obscura; on another, Agonoscelis nubila, or Migymonum cupreum; while the curious Anisocelis with leaf-like tarsi, and large species of Mutis, with immense thighs and spiked thorax, could be taken on the wing. But when so taken it was necessary to be careful in handling and disengaging them, for some of these Hemiptera are provided with a hair-like proboscis at the extremity of their elongated heads, with which they have the power of penetrating the skin and inflicting a painful sting. Such are the species of Sycanus and Eulyes, which I learned by experience to handle as carefully as if they were wasps or bees, although none that I met with were more handsome than Eulyes melanoptera, the wing-like expansions of whose body, as well as the legs, were of a rich crimson blotched with jet black. Another very beautiful species that may be mentioned as rather common is Callidra dilaticollis, with wing-cases of the richest dark green spotted with black; and several of those I brought home appear to be new to the hemipterist.

A remarkable and disagreeable circumstance, well known to dwellers in this part of the world, is that the graves of Europeans who have been buried in the island are pretty certain, sooner or later, to be rifled and desecrated by the natives; not, be it understood, by the Malays proper, but probably by the tribes of the interior. They never meddle with the graves immediately after the interment, and even years may elapse before they ultimately effect an entrance; and when they do so, it is in such a manner that it is very readily overlooked; for they do not roughly uncover the grave, but having made a small and inconspicuous aperture, they extract through it whatever the grave may contain which excites their cupidity. They rarely remove the body, though it appears the bones (simply as bones) have sometimes been found missing; but the cause of the desecration is somewhat obscure. It has been suggested to me by some, that having observed the luxurious modes of living to which the Europeans are accustomed, they believe that we must carry some valuables with us into the grave; more especially as it is the custom of their own people to bury rings, jewels, &c., with the body in their graves. But if this were the case, it would seem strange that they have not, ere this, learned the fallaciousness of the idea; for, notwithstanding that no valuables are ever buried in European graves, the desecration of them sooner or later still seems an inevitable evil.

Another suggestion has some probability also, namely, that the graves are robbed for the sake of the skull. It is said that the Dyaks of Borneo are the offenders, with whom it is a custom to collect heads, and among whom the man is great according to the number of heads he possesses. It is asserted that the heads thus taken from graves are treated like other heads, and the fiction established that such heads have been taken in fight. But there are difficulties on this theory, for the heads of their victims are usually dried while still fresh, whereas the graves are often undisturbed until long after interment, and bare skulls only can remain. If it be supposed, however, that the Dyaks, knowing the smallness of the European population in Labuan, make periodical nocturnal incursions for the purpose of taking heads from the graves, an explanation might be found in the supposition that they would search all the graves filled since the last visit, securing, however, only such heads as they might find suitable for their purposes.

Head-hunting, it is true, is now abolished in the Sarawak territory; but that part of Borneo which lies opposite to the island of Labuan is peopled by tribes which have no such scruples. Indeed the proximity of that side of Labuan occupied by the half-dozen Europeans constituting the government of the colony, to the lawless and half savage tribes of the opposite coast, easily visible at four or five miles distance, often struck me as offering singular advantages for an exterminating raid; and I have sometimes, as I lay awake on dark and stormy nights in a solitary bungalow on the sea-shore, speculated what there might be to prevent a *prahu* full of natives from landing on the beach, surprising and murdering us without the chance of resistance, and either getting back to the mainland without possibility of pursuit, after rifling the house, or carrying their extermination to the next bungalow, and indeed to all the European residences on that side. There really is nothing to prevent such a catastrophe, nor has been for the last 20 years, except the moral influence which European power has over the native mind.. There is usually, but by no means always, a gun-boat in Victoria harbour,—in reality, a perfectly ineffectual defence against a well-planned attack; but the natives having seen the resistless power of these vessels against their piratical *prahus*, have a wholesome fear of such a force; and even though no gun-boat may be in the harbour, or within a thousand miles, they have a salutary belief that one is always at hand, and within call; and, moreover, that wherever they may be, vengeance will surely follow them, and inevitably find them out.

CHAPTER XII.

LABUAN (*Continued*).

Butterflies of Labuan — Mode of Flight — Number of Species — Dominant Species — Butterflies of Pulo Daat — Hermit Crabs — Cocoa-nut Planting — Dragon-Flies — Water Beetles — Jungle Spiders — Carpenter-Bee and Mason-Wasp — Eulima and Stilifer — Alligators — Mollusca — Feather-Stars — Nudibranchs — Mantle-cutting Doris — Land-Shells — Reef at Pulo Pappan — Dendractinia — Weather at Labuan — Luminous Fungi.

ALTHOUGH the Lepidopterous insects of Labuan cannot vie with those of South America, as a rule, either in size or in beauty, there are a great number of considerable interest and of striking appearance. They are, of course, derived from the mainland, and less numerous than those of the opposite Bornean coast; but inasmuch as the jungle of Labuan is not only far more accessible than that of the opposite coast, but is also remarkably fine and luxuriant, a large number of handsome species may be obtained there with comparative ease.

The only way to capture the best species is to follow them into the jungle, although a considerable amount of skill is necessary to overcome the difficulties. The net becomes an awkward instrument in a tangled forest, and the only available method is to watch for them in small open spots, and seize upon those which pass, for pursuit is next to impossible. Many of the species fly with amazing rapidity and strength of wing, and in some cases pursue a straight line through the maze of branches, eluding nearly every attempt to capture them, except by stratagem. Others, often the most handsome insects, fly habitually so high that they are usually out of reach of the net. In all such cases the sacrifice of a single specimen will often secure others; for butterflies are gregarious, and a dead specimen pinned upon a conspicuous twig will often arrest an insect of the same species in its headlong flight, and bring it down within easy reach of the net, especially if it be of the opposite sex. Sugaring the trees has not been tried by entomologists in this part of the world; and the use of a lamp behind a sheet, found so effectual for nocturnal captures by Mr. Wallace, has not yet been seriously adopted.

The jungle-road, extending nearly across the island, and the skirts of the jungle, always proved to me the most prolific spots, the insects dashing out for a little distance and pursuing their erratic flight through the open, in which case, if near, there was a chance of a capture. But even here it was often tantalizing to see a rare or beautiful species, such as the swallow-tailed Papilio Gigon, fly out of one side of the jungle, cross the road with the speed of a race-horse, and irrecoverably disappear in the thicket on the opposite side, almost before one could draw breath. The swift flight, now over the tops of the trees, now down near the ground, was characteristic of the Pieridæ, of which Pieris andria and Callidryas alcmeone were common examples; while the Papilionidæ distinguished themselves by their strength of wing and straight headlong course. If missed by the first throw of the net no second chance was afforded, for the insects would whirl round and round the instrument two or three times and then dash off out of sight.

Another source of disappointment arose from the fact that

not unfrequently, when one thought oneself fortunate in capturing a fine insect, after carefully disentangling it from the net, its wings turned out to be so torn and rubbed as to render it almost useless, except indeed as a decoy. This circumstance is due, I imagine, partly to their frequent battles with one another, in which they whirl round each other with the greatest rapidity, and appear to be incited by the greatest ferocity, and partly to their habit of flying rapidly through the interlacing twigs and foliage of the jungle.

Certain species could always be found in particular spots; the orange and pumilow trees in the plantations always abounded with the handsome large red and black Papilio Memnon; grassy nullahs sheltered abundance of small ocellated species; the variable Papilion Pammon floated over every hedge-row, and certain bushes always harboured some swift-flying pale yellow Pieris Namouna; even a patch of sandy sea-shore generally produced a large buff insect (Cynthia arsinöe), which was fond of alighting upon it, so that, although it matched the sand well in colour, it was not difficult to secure it. But, without going into the jungle, only about half a dozen common, though handsome species, could be met with in a morning's ramble, unless, as when, by a fortunate accident, I captured a magnificent yellow satin Ornithoptera, in a pleasure-garden. Some species, too, are of crepuscular habits, and only make their appearance near sunset, when, from their large size, they might be almost mistaken for small bats. Such are Amathusia Philippus, and its allies, remarkable for the angular form of their wings.

During a month I succeeded in taking upwards of 60 species of butterflies in Labuan, a very respectable number for so brief a time, and showing considerable richness of the island in this respect. No complete collection has hitherto been made, though I have seen about 150 species in one cabinet; but a gentleman of the garrison, who has lately arrived there, after a long apprenticeship among the Lepidoptera of Malacca, is now busy with his net, and will doubtless soon make more species known. When the road above mentioned was in process of formation through the jungle, some years ago, butterflies were so abundant that they are described as having flown about in perfect clouds; and I am credibly informed, by Mr. Low, that he had taken as many as two dozen in a single sweep of the net. For the same reason, probably, cleared ground near the jungle is always most productive of butterflies; and as it is well known that wherever ground is newly cleared new plants immediately spring up, so also, under similar circumstances, a new species of butterfly is likely to occur in the first season after a clearance; but although it may be in profusion then, it does not follow that it is so in succeeding seasons; and the opportunity of securing specimens should not be lost on account of the insect appearing to be so common. As an example of this, it may be mentioned that when the compound surrounding Mr. Low's house was cleared, a beautiful species of Apatura appeared in myriads, and was abundant all that season; but ever since that time not more than one or two specimens have been observed each year.

The dominant species in Labuan are certainly Danais juventa, abundant everywhere, and Neptis aceris, to which may perhaps be added the little yellow Terias Hecabe. Danais similis is also common here; but these species appeared to be represented at Sarawak by Danais crocea. Many beautiful Papilios are met with; but perhaps the most striking and extraordinary of all the Lepidoptera are the magnificent though common moth, Nyctalemon Hector, and the remarkable clear-winged or black-spotted Hestia Lynceus.

In the immediate neighbourhood of Labuan there are two small islands : one entirely covered with jungle down to the water's edge; the other formerly jungle-grown, but now for the most part cleared for cocoa-nut plantations, but still retaining a small patch of the virgin forest. I paid a visit to the latter, and soon discovered that it abounded in the most magnificent of the jungle species; for the island, although small, is nearer to the mainland than Labuan, and, in the tangled jungle which has been allowed to remain, the most exquisite dragon-flies vie with the butterflies in beauty. Moreover, a walk under the boughs of the great trees which overhang the beach, and sometimes impede the passage at high water, gives ample employment for the net, the jungle species coming constantly out and skirting this open space. Here I was sure to meet with species, without difficulty, which in Labuan I might have searched for all day in vain, such as the beautiful species of Papilio with underwings as though inlaid with mother-of-pearl—Papilio Bathycles, and the variable P. Euripylus; or the brilliantly variegated P. Agamemnon. Here, also, a not uncommon insect was Certhosia Cyane, whose wings are elegantly scalloped and richly coloured; or the handsome Iphias Glaucippe, a large, orange-tipped species, not uncommon in China. Many long rows and sails I made to this rich locality; and thinking that the other island, Pulo Pappan, might be equally rich, I went with my net to visit it, but alas! I only saw, in all its verdurous depths, two butterflies, *both* the commonest species of Danais and Terias; and was the more vexed, on my return, to learn that the morning had been remarked in Labuan as one singularly favourable to Lepidoptera, which had been flying about in unwonted abundance. Looking for some cause for this difference between the two islands, I imagine that the first-mentioned abounding in pools of water gave it more favourable conditions for butterfly existence; while in the latter, as I could nowhere meet with a drop of water, so also I could find no Lepidopterous insects.

The time which I found most favourable for capturing these insects was from 7 A.M. to near 11; before 7 I have found scarcely any stirring, and as noon approaches they almost suddenly disappear. A few return in the afternoon, but scarcely in sufficient numbers to make a walk profitable. And, indeed, after four or five hours of such work, though not without pleasant excitement, a rest had been well earned. And then it was that a fresh cocoa-nut, added to the stores brought with us, was thoroughly appreciated; and no longer enticed from our repose by the flying gems which had hitherto allured us, I halted with my Malays on the sand beneath the spreading branches of a Dolichos, or under some shady tree festooned with epiphytic orchids. Here, listening to the ripple on the shore, and the loud song of the cicadas, and looking over the calm blue sea to the wooded shores of Labuan or Borneo, the hours of high noon were very agreeably passed, until the time arrived for resuming the net, or for a hunt upon the beach, or perhaps a sail back from whence we came.

It was at such times that I have often watched the numerous hermit crabs (Paguri and Cœnobitæ), which abound on all these sandy beaches; and where these border the jungle, they creep up among the dead leaves for a considerable distance, so that I have not unfrequently, when standing in the skirts of the jungle watching for insects, been startled by a rustle at my feet, which at first I mistook for a snake or a lizard, but on looking down I have seen nothing more dangerous than a wandering hermit among the leaves. That they eat vegetable as well as animal food I am certain, for at Enoe I saw them clustered upon mangrove shoots which had been borne there by the waves. I have often taken up one of these, to which a number of hermits clung, and even in my hands they have eagerly nibbled off the dark skin of the tender shoot. So also in breaking into old decayed trees in search of beetles, I have often been surprised to find hermits (Cœnobitæ) concealed within the heart of the decaying wood, and feeding upon it. The variety of shells which these little crabs occupy is very great, and by no means confined to the turbinated Gasteropods, as Trochus, Turbo, Natica, Neritina, &c., but Cones, Mitras, Seraphs, Turritellæ, &c., have also their tenants, in which they often appear very awkwardly and grotesquely situated. On one part of the coast of Johore, I remarked that they chiefly inhabited the shells of the thorny woodcock (Murex), which were strewed about, of all ages and sizes—some very minute and young, and in good preservation.

Lying down upon the beach, I have watched the shore-crabs leisurely crawling up the sand, and often been struck by their quicksightedness and wariness. Although perhaps a dozen yards off, if I raised myself into a sitting posture, they would instantly retrace their steps, and scuttle back; and even if I but raised my head, the gesture was not lost upon them, but they would immediately stop, and await a further demonstration on my part; but if pursued, their swiftness was such as often to elude my utmost endeavours to capture them.

The island I have alluded to as being so rich in Lepidopterous insects, is called Pulo Daat or Daat, Pulo meaning simply *island*, in the Malay language. Nor is its interest solely connected with its abundant butterflies. It is about a mile long, and contains about 600 acres. Fifteen years ago, this island was covered with virgin jungle, and was to a great extent cleared for the purpose of growing cocoa-nut palms, of which there are about 10,000 now upon it. This is a very remunerative crop; the cocoa-nuts are allowed to sprout through the husk, until the shoots are about two feet in length, they are then placed upon the soil, in which they readily take root, and grow with little trouble; but it requires a considerable period of time before an adequate, or indeed any, return can be derived from the capital invested. Daat has been planted 10 or 12 years, and as many thousands of pounds have been sunk in its cultivation; but it is only just beginning to yield what will probably turn out eventually to be a satisfactory profit. I have already stated that the proboscis monkey (Nasalis) still exists in the small jungle which remains uncleared—which is very dense, abounding in pools of fresh water, and rendered almost impenetrable by the numerous fallen stems of the Nibong Palm, upon which are dangerous long spines, arranged in close whorls, which tear the clothes and pierce the feet of the incautious rambler. The only other large animal is the wild black pig, which is pretty numerous, although very shy.

Hovering over these freshwater pools, as well as over the swampy ponds of Labuan, were always numerous very handsome dragon-flies, the most abundant of which has a bright scarlet body, and is common also at Singapore. They were all very strong and active flyers; but one very large species, with a light blue body, exceeded them all in strength and agility. These insects are remarkably wary; their habit is to fly with great rapidity over a pool of water from end to end, diverging a little from time to time, but passing over the same spot again and again. Whenever I posted myself, however, near one of these spots with my net, it almost invariably avoided coming within reach; but if I did get a cast, and partially entangled it, with a mighty struggle it freed itself, and was off like the wind; nor would it return to the same spot as long as I might wait, although I should probably find it there next day. All the other Libellulæ, however active, were to be caught, but this one evaded all my attempts.

In these pools water-beetles were not uncommon; the largest a species of Hydaticus, which were in about equal numbers with a smaller species of Dytiscus. But by far the most curious was an elegantly-shaped species pointed anteriorly, and with the borders of the wing-cases beautifully sculptured, which appears to be the Porrorhynchus marginatus of Java.

Within the jungle, one is often brought up suddenly by an immense web which entirely blocks up the way between two trees, and in which a large spider (Nephila) has its abode. This species has a rectangular body, $1\frac{1}{2}$ to 2 inches long, and very long legs, stretching $5\frac{1}{2}$ inches across, and presenting the appearance of a Longicorn beetle, its two anterior legs looking like antennæ. In most of their webs I observed small spiders, which appeared to be at home there, and probably fed upon the remnants of the larger spider's repast. One of these had a remarkable mode of feigning death when disturbed. Uniting four legs in front and four behind, they presented with the body a uniform curve, and the spider might, in this condition, be readily mistaken for a little bit of curved twig or bark.

A very large carpenter bee (Xylocopa latipes) flies about

commonly in Labuan with considerable rapidity for its heavy body, and a loud, droning hum. It is black, with a rich metallic gloss, which on the wings is of a fine purple, and the posterior legs are thickly coated with hairs. They tunnel into posts and other wooden substances, where they construct cells in which they deposit their larvæ, supplying them with a farinaceous paste of pollen, which they brush off and collect by means of their hairy legs. Another Hymenopterous insect of large size, and whose habits were very interesting, was a species of Sphex. I watched this insect construct its clay cell upon the back of a window-shutter in the verandah. Having brought some moist clay in its mouth, it daubed it in a circular form upon the wood, and returning frequently with fresh mud, it completed the cell in about two hours. When finished this clay cradle was about $1\frac{1}{2}$ inch in diameter, and about $1\frac{1}{4}$ high. As it came back repeatedly with a fresh stock of clay, it was amusing to see it search for its chosen site. There were several shutters in the verandah, all very much alike, and which should properly have been close back against the wall, but some of them were about a foot or eighteen inches from the wall, and this one was so placed. The Sphex would fly into the verandah after an absence of a few minutes, and try several shutters before he came to the right one upon which the cell was building. It is the habit of these insects to deposit their eggs in this cell, placing therein also some disabled caterpillars or grubs which cannot escape, so that the larvæ, when hatched, at once find a ready supply of food. The wasp most frequently met with in the jungle was a moderate-sized species (Bembex melancholia), with the abdomen banded with a metallic blue and black. In sandy spots a large and very long-waisted insect, of very venomous aspect, scooped holes in the ground, and warned the entomologist to be careful in walking over a mine of stinging wasps. This was Eumenes circinalis, and a true wasp (Vespa cincta) also commonly occurred. A little Trigona was often caught in the net, but in the jungle these Hymenopterous insects are rather dreaded, as it is not an uncommon thing to disturb a nest of bees or wasps, which, thus alarmed, sting *ad libitum;* nor is it easy to avoid their weapons, a hasty retreat being of little avail, unless a pool is near, an immersion in which is the best protection.

Many other interesting species of insects were met with here, such as Mantises, which might be usually obtained by sweeping the long grass with the net; or the curious walking-stick insects (Phasma), and among the various locusts the large leaf-winged Platyphyllum.

Upon a common species of Asterias (star-fish) lying half-buried in the sand of the Labuan shore, I found numerous minute shells, which I supposed at the time to be Stilifers. They were of two species, and of the numerous individual star-fishes nearly every one had some of these little mollusks upon it. One was a slender, dark-coloured species, and usually made its appearance upon the dorsal surface of the Asterias; and the second was a stouter and larger pale species, principally found in the angles of the arms or rays, or upon the under surface. Few star-fishes were entirely free from them; such were, in fact, the exceptions. Some had three or four dark, and four or five pale specimens upon them, while others had only one or perhaps two. Maimed star-fishes, of which there were many, having lost one or more of their rays, usually had no shells upon them; but young, small-sized specimens in nearly every case possessed good specimens of the pale species. On examination of these little mollusks, however, they proved to be species of Eulima. The shells were very transparent and delicate, and the body of the animal could be distinctly seen through them with a lens; the black eyes also, surrounded with yellow irides, were easily seen through the shell.

The star-fishes, when taken up for examination, usually ejected a jet of water from the centre of the dorsal surface, the serrated edges of the plates of which opened beside the tubercle; little jets of water also spurted out from the extremities of each of the rays.

Two or three species of Holothuria were met with; but one, a large black one, abounds here and in most other places. This is the Trepang of commerce, and is collected largely by the Chinese, and dried and eaten by them. Upon them I usually found Stilifers in the neighbourhood of the gill-tufts. On being touched they emit, with great violence, a large mass of tenacious, bluish threads, which stick to the hand with such adhesive force that it is difficult to rub them off, but no irritation ensues. Holothuriæ of any species are difficult to preserve alive, since they usually eviscerate themselves on the first night after their capture. The Stilifers, however, are more readily discovered when the animal is dead, as, during life, the contraction of the orifice draws them in out of sight.

With regard to the other marine productions of Labuan, two or three rocky reefs running out on the east side of the island gave me many opportunities of searching for littoral animals at low tide; and although my searches failed to discover some species which I was particularly anxious to meet with, there was plenty to reward patient investigation. One could not help feeling that there was a certain amount of risk in wading nearly up to the neck in these waters, and the information I received concerning alligators did not make me feel more secure. One of the residents assured me he had frequently seen alligators in the bay as he rode by on horseback, their noses just appearing above water; and it is in the records of the place, that not long since a man, who went down at night to wash his rice-tin at the water's edge, was carried off by one of these monsters. On another occasion an alligator seized a woman near this spot, but her dress getting round its head impeded its movements, so that the cries of the woman bringing assistance, it was seized and killed. The long extremity of its nose had been broken off in a former encounter. Another alligator killed here was found to have a digested ball in his stomach, consisting of the body of a man, the bones all broken, and the clothes all rolled up with the flesh in a scarcely distinguishable mass. Happily I never was troubled by a visit from these unceremonious gentry, though the thoughts of them seldom left my mind quite free.

The Mollusca found on these reefs are not numerous, although a great many species may be obtained by making excursions from Labuan, as a central point, to the various small islands within reach. These include no less than 37 Cones and 36 species of the beautiful genus Cypræa. None, however, of the more rare and valuable species are included in this list, though several are highly interesting. Olives crawl about the sand, leaving tracks by which it is easy to discover their hiding-places; they are of several species, the most common being Oliva acuminata and O. maura. In a part of the coast where a small stream runs out, a black, muddy patch is formed, occupied by hundreds of holes of Gelasimi. It is an uninviting-looking spot, but a number of Olives are produced here of a very rich dark brown approaching to black. This is probably a local variety.

Every stone is covered with tunicates and sponges of various forms and colours, and beautiful silky worms occur, whose delicate lateral fringes run into the skin upon the slightest contact, causing considerable and disagreeable irritation. Echinoderms, however, did not appear to be abundant either here or in any other of the places which I visited.

Some magnificent Feather-stars (Comatulæ) presented themselves, which I much regretted could not be preserved in some way. Their forms were so complicated, however, that time did not permit of my drawing them with such accuracy as to be useful for scientific purposes, and I was not successful in preserving them entire. One of them was of a rich carmine, and $6\frac{1}{2}$ inches in diameter, breaking very easily; and even in the fresh sea-water it discharged its colour very rapidly, pouring it out like blood, staining the hand, and strongly tinging the water; and itself passing from carmine to a rich yellow, and thence to crimson, until nothing was left but a quantity of dingy fragments, which gave no indication of what it had been. A second species was of a rich olive green, with the distal ends of the arms white for $1\frac{1}{4}$ inch, the whole diameter being $9\frac{1}{2}$ inches. This splendid specimen did not show any special inclination to break up, but discharged its colour to some extent into the water; and in the attempt to dry it, lost it all.

I have found the directions given for killing these animals entire quite ineffectual. In the case of a Comatula which came up on the anchor in Haitan Straits, I was successful in preserving a record of it, by drawing and description; but

Group of Nudibranchs from the China Sea.

Localities (from left to right): 1. Ke-lung; 2. Haitan Straits; 3. Ma-kung; 4. Labuan; 5. Raleigh Rock; 6. Fiery Cross Reef.

To face Page 195.

upon adding the minute quantity of corrosive sublimate, the animal rapidly discharged its colour, and broke up into minute fragments. This species also showed no inclination to break itself up while in health, even when handled; and, indeed, this peculiarity, usually supposed so characteristic of the feather-stars, is by no means universal. Two large-sized and remarkably beautiful specimens of different species, obtained at Singapore, I handled with impunity without breaking the smallest portion of them; but, unfortunately (and this well illustrates the difficulty of doing all one would wish, even under apparently advantageous circumstances), these specimens were obtained so late in the day that it was impossible to do anything with them till daylight reappeared. But, alas! the bucket which in the evening contained two healthy and splendid feather-stars, held in the morning only an offensive mass of small fragments, the colour of the water and of the remains being equally unattractive.

Nudibranchs here appeared to be few, or my ill-fortune prevented me from discovering them. Mr. Low told me that he had frequently seen very beautiful species, of which, however, he had not taken any particular note, and kindly took me to where he thought I should find them, but we both were equally unsuccessful. The first I met with I at once recognised as the blue Doris (D. Barnardii) of Makung Harbour. Next time I obtained the crimson-spotted one already obtained at Slut Island in Haitan Straits. I was ultimately successful in getting a very elegant species, striated along the back with delicate alternate lines of deep brown and yellow. This appeared to be a not uncommon species on these shores, and I have met with it nowhere else. Some very beautiful Planarian worms, which at first had the appearance of nudibranchs, and were not less interesting, occurred from time to time under the stones; but these were not numerous in species.

Besides the reefs of Labuan itself, each of the three small islands between it and the mainland had their special points of interest. At Daat several specimens of a large Dorid occurred which was not found elsewhere. They were about four inches long, and two inches wide, with expansive gill-tufts and large tentacles; but not such beautiful animals as most of their tribe. They had something of the appearance of Doris tuberculata, though less variously coloured, being usually grey and studded with tubercles above, the whole under-surface smooth and blotched with irregular black spots of various sizes upon a grey ground. I carried home specimens of this animal for drawing and examination; but on looking at them the following morning, I found the wide projecting margin of the mantle cut off close to the foot, as though with a sharp scissors, leaving the thick slug-like body, which rapidly decayed. At first I could not understand what had done this, but immediately afterward a second specimen performed the same feat. Attributing this spontaneous amputation to a suicidal act, arising from the fouling of the water in which they were placed with some other animals, and wishing to preserve the last specimen, I placed it by itself in a large vessel with fresh salt-water; but next morning I found it severed like the others.

The adjoining island (Pulo Pappan), though barren in Lepidoptera, offered features of great interest in other respects. In the thick jungle land-shells were tolerably abundant; two or three elegant Cyclostomas being found on the under sides of the leaves, and two species of Helix of great beauty, tolerably common. One of these, Helix atrofusca, of various delicate shades of light brown, was in considerable numbers, while the other, Helix læta, of a pale straw-colour, was less numerous. The magnificent land-shell, Helix Brookei, only equalled by a species in Cambodia (which indeed surpasses it), is pretty frequent not only in Borneo, but in Labuan, as may be estimated by the fact that, although once a high-priced shell, the natives who are employed to collect, only ask ten cents (5d.) each for them at the present time.

It was on this reef of Pappan that, in company with Mr. Low, we found the great Anemones which sheltered fish, and which I had previously observed on Fiery Cross Reef. It has long been known that a sea-cucumber (Holothuria ananas) shelters a fish in a similar manner, and a figure of this fish is given in the "Voyage of the Astrolabe." It is not a little remarkable that a Holothuria of the same habit exists on this reef, as has been discovered by Dr. Coulthard of Coal Point; but whether it is the Holothuria ananas or another species I am not aware. Beautiful living corals strew this corner of Pappan; Fungiæ, with large bright green club-shaped tentacles, looking like magnificent anemones; patches of large dark purple Echini, containing from 50 to 100 individuals, so closely packed as completely to conceal the sea bottom; their spines six inches long, and rows of brilliant metallic blue spots glistening in the ambulacra; little Asterinas (A. minuta?), and numerous shells of Murex and Cypræa. Elegant little Gorgoniæ grew up here and there, inviting a hand to pluck them; but woe to the hand that accepted the invitation, as one of our Malays did, for the ramifying stem was covered with a transparent gelatinous substance which stung him just as the threads of a jelly-fish or sea-nettle would have done, much to his astonishment and discomfiture. I was somewhat surprised to find *under* the stones in the water some Peroniæ, those erratic slug-like creatures which usually creep about on the rocks above high-water mark; and lastly, the peacock-tailed seaweed (Padina), which I had found in the Pescadores, in North Formosa, in the adjacent islands, and in many other places, was here growing in great profusion.

A third island was Pulo Enoe, the most southerly of the group, and a mere clump of trees, connected to Labuan by a reef, upon which were numerous hammer-muscles (Malleus), Pinnæ, and other shells. But one animal from the reef was of far more interest than all the rest. This was a magnificent species of anemone, which has its abode in crevices of the rock, just below low-water mark. I succeeded in removing one entire, which gave me an opportunity of recording this beautiful new genus. The tentacles instead of being simple or club-shaped, as usual in most species, were singularly ramified, each tentacle giving origin to several branches, and each branch terminating in a fine curved tendril-like branchlet. This beautiful sea-flower was seated upon a thick corrugated yellow column, and measured, when expanded, nearly five inches across, the ramifications of the tentacles being picked out with a bright yellow line, and the central part of the disk brownish pink. The Actinia, described by Quoy and Gaimard under the name of Actinodendron, bears no resemblance to this species, and yet no fitter name could be given to the genus than Dendractinia. I afterwards found a similar individual on the beach west of Singapore, but was unable to secure it.

Nearly every evening (in August—September) during my stay in Labuan, the sky clouded over a little before sunset, and became gloomy, and during the night it often rained heavily. It is said that 160 inches of rain per ann. fall in the island,

and I was assured by an old resident that he had known seven inches fall in a single night. There was also thunder and lightning nearly every evening, seldom however coming quite close, but usually hanging over the mainland, though Pulo Daat seemed to come in for a good share of rain and storm. But, however gloomy and wet the night may have been, the mornings, as a rule, were beautifully fine, and the air from 6 to 9 A.M. delightfully cool and pleasant. After that the sun gained power very rapidly. On the 28th of August, a little before sunset, the sky having its characteristic hazy appearance, I observed very distinctly the phenomenon of parhelia, or mock-suns—a mock-sun being on either side of the real luminary, and indistinguishable from it in point of brightness.

With so much wet it might have been expected that fungi should abound in the jungle. They were not numerous, however, though there was one of considerable interest. I had observed, on passing a plantation late in the evening, numerous patches of faint light scattered over the ground and upon the grass; and picking my way cautiously over stumps and ditches towards it, I found that the light proceeded from a fungus growing upon the tree roots, generally, but not always, on old and decaying ones. The fungus is a species of Agaricus, and is pronounced indeed to be the A. Gardneri which grows in Brazil. It shone with a distinct but pale light, very soft, and of a pale greenish colour, the young specimens appearing to give a more intense light than the older ones. It was of a cream-colour, thin, soft, and fragile, with the texture of a Helvella, and with white spores. On visiting the spot next day those which had appeared to be fresh and young on the preceding evening were becoming brown, and apparently decaying; so that they are probably very short lived, perhaps only lasting in perfection one single night, and then replaced by others. They seem, moreover, to be variable in quantities and brilliancy; and the following night I could only meet with one or two small specimens in the same plantation.

Malay Houses at the Anchorage, Sarawak.

CHAPTER XIII.

SARAWAK.

Entrance to River—Antimony Anchorage—Tarnuh-puti—Drift-wood—Town of Kuching—Former condition of Sarawak—Sir James Brooke—Prospects of the Settlement—The Tuan Muda—The Dyaks; their Superstitions—Miss Burdett Coutts' Estate — Gambier planting — Flying Squirrel — Flying Lizard—Flying Foxes—Vegetable Productions—Rain—Italian Naturalists —Quadrupeds—Domestic Animals—Dyed Fowls—Reef at Pulo Barundum.

On the 17th September my jungle rambles came to an end, and we set sail for the Sarawak River, which we began to approach on the 21st. The entrance is marked by the bold promontory of Tanjong Po, a wooded limestone headland with sandy bays at its foot; while further westward rises a yet higher point, Tanjong Sipang,—the two including a mountainous peninsula, upon which Mount Santubon, 2712 feet high, forms a landmark for the westerly entrance of the river. To the eastward the land is low, consisting of mangrove swamps and low jungle, the tops of the trees only being visible. Guided by a buoy placed by the Sarawak Government, the bar is crossed, and having rounded Tanjong Po, the river's banks assume for the remainder of the way to Sarawak a uniform and somewhat tame appearance. The shores are low and muddy, clothed with a beautiful vegetation, consisting of a fringe of Nipa palm-trees (Nipa fruticans), with pinnate feathery leaves, 20 feet long, arising from the ground without any stem; and these are intermingled with low jungle trees and bushes, washed at the base by the flowing stream, and affording in their muddy creeks and hollows appropriate lairs for alligators. With the exception of a Malay hut here and there, there is scarcely any variation in the scenery, nor much life visible, except now and then a large kingfisher, or hornbill, or some long-tailed monkeys chasing one another in the trees.

At the distance of 17 miles from Tanjong Po the river divides, and beyond this ships of any considerable draft seldom pass. This is the Antimony Anchorage, so called from the antimony stores of the Borneo Company, which are located here. Here were several square-rigged vessels; and as we proceeded we met many canoes, containing two or more Malays, who rested on their paddles to watch us as we passed. Some larger craft also there were, having an European build (lorchas), and flying the Rajah's flag—a broad cross, half red and half black, upon a buff ground. As we neared Sarawak, houses became somewhat more frequent in gaps or clearances by the river side, built upon piles over the mud, and covered with attaps. Notices posted up at intervals, such as "Rocks, hug this shore," indicate the difficulties and intricacies of this navigation, and the leadsman sometimes called seven fathoms although only about a boat's length from the bank. It was here that the "Samarang" touched the rocks and was thrown on her beam-ends, only recovering her position after incredible labour and considerable time had been spent upon this apparently hopeless task.

Near Sarawak is the straggling village of Tarnuh-puti (white earth), a brick-making place; and it was not a little amusing to see the groups of men, women, and children squatting on their hams, or gathered together in knots, discussing the unusual sight of a large ship so high up the river,—women with the sarong fastened under the breasts, the children for the most part unencumbered with any clothing whatever, their round and open eyes expressing not a little bewilderment. The weather was cool, and the useful sarong was turned into a cloak, a hood, a comforter, or what not, as occasion required or caprice suggested; and the groups of curious faces, and swarthy forms, would each have well repaid the trouble of a separate photograph.

Immediately afterwards Sarawak appears in sight, the Malay houses extend along the river's banks for a considerable distance; and opposite these houses we cast anchor in 10 fathoms' water, although there was but just width for the ship to swing. An immense collection of drift is brought down by the muddy waters; and when the tide turned we were surrounded by a quantity of broken trunks, old logs, long leaves, and sometimes whole trees of the Nipa palm, seeds of screw-pines (Pandanus), and *débris* of all kinds, which oscillated backwards and forwards all day, and often caused some inconvenience by getting entangled in the tackle of the ship.

Sarawak consists of a long line of Malay huts, built on piles on the left bank of the river, broken midway by some rising ground, upon which the houses of the Rajah and residents, with their compounds, are situated, and occupying a bend of the river. On the right side are the Chinese town and bazaar, Kling quarters, barracks, the fort with a six-gun battery, Government offices, a sago manufactory, the Borneo Company's establishment, and some few European residences on the hilly ground behind. In the Chinese bazaar may be purchased many European articles, such as Rimmel's scent, eau de Cologne, dolls, &c., as well as many common articles of crockery,* and ornamental ware, similar to those which may be found in cottages in England. It is a busy, lively quarter on the river side, where numerous boats are constantly passing and repassing, and passengers landing and embarking; while a miscellaneous throng of Chinese, Madrasees, Malays, may generally be seen, to which is added sometimes a small party of Dyaks. Behind the town at some miles distance rise several lofty peaks, which form a picturesque background, of which the chief are Matang, Singhi, and Peninjau.

A visitor to this town of Sarawak, or, as it is called by the Malays, *Kuching* (a cat), who may have chanced to know anything of its state a quarter of a century back, may well be struck with its flourishing condition, and with the aspect of peace, plenty, and security, which now pervades a place so short a time back a prey to lawlessness, rapine, and bloodshed. No portion of the globe could have been more

* It is not a little curious that I saw here some plates on which the time-honoured willow-pattern was *Anglicised*. There were the bridge, the men, the birds and the trees, but all stripped of their Chinese features and rendered in English. Where they were made I do not know, but I never saw them in England.

wretched than this territory thirty years ago, when pirates and robbers swept the country with fire and sword; when murderous head-hunters sought for their bleeding trophies far and near; when savage tribes sought opportunities of making a raid upon the least protected of their neighbours, murdering all the males, and leading the women into captivity. Such was the reign of terror, and worse than civil war, which Sir James Brooke found existing in this part of Borneo. Far from the seat of even nominal government, the strong hand kept down the weak with the ferocity of the savage, and without appeal; and as a necessary result, the country was rapidly becoming depopulated; for those who escaped the *kris* of the enemy could only look to die of starvation. But the philanthropy of Brooke was not content to pity the unfortunates, in whom his penetration saw traits of character and capabilities of improvement, which events have fully borne out. First, having with a superhuman effort given such a blow to piracy that it has never been able to lift up its head since, and having fairly scotched, if not killed, the snake, he thus essentially mitigated the great crying evil of that part of the world, and paved the way for negotiations, which the natives readily appreciated and soon sensibly adopted. Having shown himself *fortis in re*, he next exhibited his character of *suavis in modo*, and easily succeeded in winning the entire confidence of the population, and by his own indomitable will and enthusiastic nature, backed by no state support or military force, he has changed the desolated district into a thriving settlement, well governed and secure, where every man sits under his own vine and under his own fig-tree, none daring to make him afraid. Associating with him the hereditary native chiefs, he has banished all jealousy of foreign rule, and has endeared himself and his name, and the English nation, to the people he has so worthily governed.

The territory of Sarawak is magnificently watered and very fertile; and that a flourishing trade of any description should so speedily occupy the place of lawlessness and plunder is a surprising phenomenon. Upwards of a quarter of a million sterling of exports and imports now pass along the river annually, and that they are not more valuable can hardly be wondered at, when we consider that up to the present time the territory has been, as it were, a private appanage, and unprotected and unrecognised by any western power. That it is *not* is no fault of Sir James Brooke; and it may be hoped that our Government will not permit a ready-made and valuable colony to pass into the hands of any foreign power, however desirous such foreign power may be of making so desirable an acquisition.

It is to be feared that Sir James Brooke's career of usefulness is over, and that he will be physically incapacitated from returning to the scene of his labours and peaceful conquests. He leaves, however, as his recognised successor, his nephew and adopted son, Charles Johnson Brooke, who is styled Tuan Muda, and who has completely gained the confidence of the people over whom he is placed. This gentleman lately published a work entitled, "Ten Years in Sarawak," in which he tells, with simplicity and straightforwardness, of the expeditions he has made at various times for the establishment of peace and security within the territory. By the coolness and determination exhibited in these expeditions, and by other means, he has acquired an extraordinary influence over the Sea Dyaks, who universally regard him as their ruler and head.

The Dyaks or aboriginal tribes are divided into Sea

Dyaks and Land Dyaks, but the former derive their name solely from the fact that they are accustomed to the sea, and to marine expeditions, and it does not imply that they all live upon the coast. Indeed they penetrate into the interior equally with the others; but from their enterprising and wandering habits they have naturally more force of character than the more stay-at-home section. The Sea Dyaks were formerly the great pirates and head-hunters, as might be expected, and are now a reformed class, who possess qualities which are at once of a fiery and impressible nature. The Land Dyaks are those which inhabit the upper part of the Sadong, Sarawak, and Samarahan rivers, but are fewer in number, and less warlike than the others. These two races, however, differ radically in their language from each other; and indeed the Dyaks of different tribes are often unable to understand one another, although there is doubtless an affinity between them. They have, however, no written language, as I was assured by the Bishop of Labuan, who told me he had searched diligently without finding any trace of such.

I must leave for others, however, a history of the Dyaks of Sarawak. Some account of a visit to their homes will be found in the following chapter, and I will only now mention one circumstance which occurred while I was there. Like other uncivilised nations they are very superstitious, and their superstitions are often of a very childish nature. One day when I was dining with the Tuan Muda, a Dyak came over from Sadong expressly to inform him that they had discovered a *Hantu* or bogey, in which, they conceived, he would be greatly interested. They said that the Hantu had taken up his abode in a Dyak-house, and described him as sitting down, his legs very long, and his knees reaching above his head, which was covered with long, white hair. The Dyaks fed him, and the Hantu greedily devoured all that was offered to him; but in return for their good treatment, the only words he could be brought to utter were those equivalent to "You are a fool!" The description sounds like that of a harmless lunatic, it must be confessed, and was not sufficiently interesting to induce a visit to the spot.

When relating the circumstance, Mr. Brooke told me that three years ago some Dyaks came from a considerable distance, bringing with them for his inspection what they were pleased to designate a Hantu, carefully wrapped up in a piece of cloth. They had walked for three days through the jungle, and had abstained from speaking to any one by the way, full of the importance of their mission. When they had arrived, they ceremoniously laid their treasure before him, when lo! the mysterious wonder disclosed to the Tuan Muda's eyes was — a pumpkin or gourd, dried and blackened with smoke, and having on the top the half of a cocoa-nut shell, with some of the fibres hanging like scanty hair upon it. It had been in the possession of the Dyaks for many generations, and they regarded it as a charm of the greatest potency, and in bringing it for the inspection of their chief, they exhibited to him the highest feelings of respect and regard. The Tuan Muda, however, unwilling to hurt their feelings, and respecting their motives, examined it gravely, and pronounced it to be truly a great curiosity, dismissing them with a dollar or two for their trouble, whereupon they packed up their Hantu and returned with it whence they came. Had he wished to purchase it, they would have demanded an enormous price, as they set great store by it as one of their most precious valuables. What would they have said to a mandrake?

Near Tarnuh Putih is a plantation belonging to Miss Burdett Coutts, who, it is well known, has spent a considerable sum of money in the settlement. Among other benevolent schemes she was anxious to give employment to a number of natives in cultivating the soil, and for that purpose engaged an agent to purchase a piece of land from the Government. It was doubtless her intention to have benefited the Dyaks, and for that end the plantation should have been at some considerable distance on the landward side of Sarawak. But by some strange mismanagement the land was taken on the side of Sarawak most distant from the aboriginal tribes, and where it could by no possibility be of any advantage to the Dyak population. Added to this, the spot selected is most unfavourable for cultivation, the soil being nothing better than a poor and unproductive sand. A new agent succeeded to the management as soon as the location was fairly and irrevocably fixed, and to him was left the unremunerative duty of making the most of a bad bargain. This gentleman, Mr. Martin, long a resident in Java, has gone to the work with a good will, and has undoubtedly done all that could be effected under such disadvantages. During the two years of his management this spot has been cleared; and finding that Bananas were scarce and high-priced in the Bazaar, he determined to plant that tree. It has succeeded well; but the cunning natives immediately followed suit, and Bananas at once became as abundant and cheap as they were previously scarce and costly. The success of the Banana crop, however, has encouraged Mr. Martin to try others, and he is now cultivating pepper plants.

In the neighbourhood of this estate is a Gambier plantation, the only one in Sarawak, which I visited with Mr. Martin. At that gentleman's instance a Chinese planter came forward, and he also induced the Rajah to offer encouragement to such as would promote the interest and industry of the settlement in that manner, for hitherto no attempt has been made on the part of the Sarawak Government to improve the land, or to encourage enterprise of that kind, by which the resources of the territory or its exports might be increased. The planter in this instance had the land free of charge, with the proviso that he should clear it, and plant it with Gambier (Uncaria Gambir). In the midst of the clearing, a lofty building was in course of erection, in which the processes of boiling the leaves and preparing the extract were to be carried on; and the Gambier plants were springing up healthily among the stumps of the forest-trees—some in flower and some in seed; in the latter case, resembling in appearance and contents the long pod and feathery seed of Epilobium. The process of clearing the jungle is a gradual one—all the wood being valuable for the purposes of fuel in boiling the gambier—and as it is consumed in proportion to the quantity of the latter produced, year by year the cleared space increases in extent. Gambier is a native production of the place.

Sitting in the verandah of Mr. Martin's house about sunset, I had an opportunity of observing the habits of the flying squirrel (Galeopithecus), the *Kubong* of the Malays. The animal came streaming through the air from a distant clump of trees, its flank membranes extended, and its long tail stretched out behind, and with a graceful sailing motion at length arrived at a tall tree trunk which had been left in the midst of the cleared jungle, on the lower part of which it alighted. The animal then began to ascend the trunk in a spiral direction, running a little way at a time, and then stopping. Having reached the branches, it selected one, along which it crept until it had reached the extremity, when it suddenly launched itself into the air, and glided away on outstretched wings, in the direction of another tall tree about 150 yards distant, gradually *descending* as it proceeded, and finally alighting upon the lower third of the trunk. Again it crept up to the branches, and again it cast itself off—making this time for a more distant tree, when it was lost to view in the jungle. At the same moment, another Galeopithecus arrived at the first-mentioned tree, which, standing alone, offered a good mark, and a convenient resting-place for these singular animals. This one repeated the same process, only going in the opposite direction. Every evening at the same hour these animals, probably the same individuals, might be seen making use of the same trees in their flight, so that it was easy to say when they had alighted anywhere, what would be their next flight. Having reached the highest part of the tree, they sailed steadily away to the next with grace and swiftness, in a gradually falling line, with no apparent movement of their flank-membranes, but with the evident power of accurately guiding their flight to the next stage in their progress, which may thus be described as a vertical zig-zag. The skins of these animals are much valued, and they are very abundant in many places. Coal Point, in Labuan, is called by the Malays Tanjong Kubong, or the Cape of Flying Squirrels, from the number of them which formerly existed there; but since the cutting down of the jungle in the progress of the works, they have very materially diminished in numbers.

Near the same spot, in the heat of the day, I saw the little flying lizard (Draco volans) alight upon a tree by the road-side. It flew quickly along, and straight, like a bird, without any butterfly-like fluttering, and suddenly settled upon the bark just as a creeper (Certhia) would do, for which at the first moment I mistook it. Then it ran a little way up the trunk in a spiral direction, and presently stopped to look at me. I approached in order to watch it, when the little creature stood still, and twisting its head completely round, regarded me with a stare, while its little conical pouch, which hung flaccid beneath the throat, was from time to time momentarily distended, assuming a semi-crescentic form, pointing forward in a menacing manner, and then falling again. I clapped my hands, and tried to make it fly, that I might observe its movements, but it remained looking at me imperturbably; and although I threw sticks and stones up, it only ran a little higher up, and then stopped and watched me again. The heat was so intense, that I was fain to go on my way—and none too soon—for I found afterwards, to my cost, that I was at that moment qualifying myself for an attack of fever. I saw the little Draco again in the interior, and afterwards in the neighbourhood of Singapore.

Every evening, about sunset, on the Sarawak River, the air was alive with large bats or flying foxes (Pteropus), called by the natives Kalongs. They began to appear as nearly as possible at the same minute every evening—a few stragglers first, gradually increasing in numbers, until, in the course of a quarter of an hour or so, they might be seen all over the sky, flying just out of gunshot range, but all bound in the same direction, viz. from N. E. to S. W. They flew with a heavy, slow, and steady flight, and might easily have been mistaken by a casual observer for rooks returning to their nests. The body was heavier than that of a rook, however, and there was a peculiar bat-like form of wing which at once arrested attention when they were directly over-head. They had all passed over before it was too dark to see them, and returned again before the sun had fairly risen next morning. These large bats are all frugivorous, and were bound to the fruit districts, where they spend the night in feasting. The distance which they flew to their feeding-ground must have been considerable, but they appear well calculated for long flights, having none of the vacillating and fluttering motion which characterises the insectivorous bats (popularly called Flittermice) of this country, which take their prey upon the wing. They fly at varying heights, according to the season and weather. If fine, they are usually out of gunshot range, but at other times it is not difficult to wing them. When thus brought down, they are very pugnacious, and bite fiercely; and in Java it is a common sport to match a terrier against them, when brought to the ground.

In the Straits of Banca, a considerable number of Pteropi flew across at sunset from the Island of Banca to the Island of Sumatra (north to south). They were of a larger size than the Sarawak specimens, and flew in many cases within gunshot. The two legs projecting slightly behind, gave them the appearance of having a forked tail.

The fertility of Sarawak is mainly due to an original rich loamy soil, which has for ages supported a succession of forests whose decay has produced a deep layer of vegetable mould—so that in most parts it is easily cultivated, and gives an ample return. The Dyaks, whose agricultural operations are of a very primitive character, nevertheless cultivate extensive gardens—more particularly along the river's banks, in which fruit-trees are the principal product. The sugar-cane is, however, frequently one of their contents, inasmuch as it grows luxuriantly with little or no attention, and is at the same time a very favourite article of consumption, no less with the Dyaks than with the Chinese —both of whom chew the raw cane for the sake of the juice. The produce of these gardens, which is disposed of in the Bazaar at Sarawak, is probably sufficient for the maintenance of the Dyaks possessing them, for their wants are few, and scarcely go beyond the means of subsistence; their chief food is rice, and $4l.$ or $5l.$ sterling per annum will supply rice enough for the consumption of a whole family. The cocoa-nuts also, which abound, yield not only plenty of food, but are useful for a hundred different purposes. They are to these people what the date-palms are to the inhabitants of the African coast. But perhaps the bamboo (Bambusa arundinacea) exceeds all other trees in economic value—for not only does it afford an article of wholesome food in the young shoots, but a thousand things are made from it—an enumeration of which would take up too much space; but the traveller meets with it at every turn, and the most ingenious adaptations are constantly arresting his attention, in which the joints, or septa, often play an important part. The cane itself is one of the most ornamental trees of the tropics, and here reaches a height of 60 feet, the strength and elasticity of its wood being unsurpassed. In these uses also the rattan (Calamus Rotang) shares in a minor degree.

Even the stemless Nipas (Nipa fruticans) have important uses. From the leaves—often 20 feet long—are made those useful and readily-applied *attaps*, which form the roofs for native houses; a preserve is made of the fruit, and salt is extracted from the burnt leaves; while the cylindrical and shapely stems of the Nibong are turned into ready-made posts, upon which their slight houses securely rest.

Many useful trees also grow in the jungle, the nature of which is no secret to the natives; and dammar-resin, sago, vegetable tallow, malacca-canes, rattans, ebony, camphor, and rice, are among the substances which the territory of Sarawak sends out in exchange for the silks, Javanese handkerchiefs, European cloths, China-ware, brass wire, and cooking vessels, salt, and opium, which are used or consumed therein; while to these vegetable substances, accumulated by the industry of the people, must be added its mineral treasures—antimony, gold, and diamonds; and certain animal productions, as birds'-nests, sharks'-fins, tortoise-shell, bees'-wax, and salt-fish.

In such a well-wooded country there is no lack of rain, which pours down in incredible torrents—of which I was a witness; and where high land aids in its formation, I can have little hesitation in accepting the estimate of one who by long residence was qualified to judge, that as much as 300 inches of rain fall annually in some parts of Borneo. This, with the aid of the freshets which come down the river, produces a rich alluvial soil along the banks of the Sarawak and other rivers, where the most beautiful flowers may be seen—sweet-smelling Clerodendrons and Bignonias, beautiful Cinchonaceous plants, such as richly-coloured Ixoras, purple Hoyas, and long-stamened Combretums; and a wonderful variety of those singular botanical phenomena, the pitchers of Nepenthes. Mr. Low, who has made several interesting excursions into the interior, told me that on the Limbang River, just north of the Sarawak country, he met with a new Rhizanth, allied to Rafflesia, but smaller than R. Arnoldi. He preserved and kept this plant for some years, but did not I believe make it public, and ultimately, but quite recently, hearing that an Italian botanist was making collections at Sarawak, he sent him the specimen.

The gentleman alluded to, Signor Beccari, has been for many months established upon the hill called Matang, near the town of Kuching, which he makes his head-quarters for botanical expeditions, conducted with great zeal and diligence, and from which valuable results may be expected. Another Italian naturalist, the Marquis Doria, has also been residing some months at Sarawak, employing a staff of assistants for collecting insects, birds, quadrupeds, and land-shells, from all the country round. He had naturally succeeded in amassing a considerable amount of material, which will doubtless—as well as Beccari's collections—form interesting matter for the publications of the Florentine Academy. The Marquis Doria had quitted Sarawak just before my brief visit.

It has been remarked that large quadrupeds bear no proportion to the luxuriance of vegetation of the tropics, and the greatest herbivorous animals abound most where the soil and climatal conditions do not encourage the greatest development of vegetation. Thus it is in Borneo, where, although the country is a vast forest under an equatorial sun, large animals are rarely met with. For although it is said that elephants are found in the north-west part, a general and well-founded opinion prevails that they are not indigenous to the island, but have been derived from animals imported from India for purposes of luxury or display, which have been allowed freedom to save their owners the pains of caring for so huge a beast. There can be little doubt, however, of their presence, although they may be rare; but the country is well adapted for them, and, once introduced, there does not appear any reason why they should not flourish. Deer and wild cattle are almost the only other known quadrupeds of any considerable size; for, although there are numerous small species, no large feline animals exist in Borneo, the most considerable being a species of arboreal panther (Felis macrocelis). Pigs monopolise the forest as far as the earth is concerned, and in their passage across rivers form an abundant supply of food to the alligators, which seem particularly partial to pork, probably because most easily obtained.

But the animals which are most characteristic of this part of the world are the Quadrumana (monkeys). Being essentially arboreal in their habits, they have flourished freely, and their species have multiplied indefinitely; so that a vast variety of them inhabit the Bornean jungles, many of which are probably undescribed. Among them the largest are probably not inferior to the great equatorial African apes in size, and of these orangs or mias there are two species, while most other old-world families have their representatives here.

The only domesticated animals which the Dyaks keep are dogs and fowls. The dogs are small, prick-eared, and sharp-snouted. They do not bark like our dogs; but at the same time they cannot be said to be voiceless, for they howl most musically. Their fowls are of a small breed, and have not yet been improved by any admixture of races from Europe. Geese and ducks do not trouble the Dyaks, though the latter are kept by the Malays. In a street of Kuching I observed a fowl which arrested my attention by its rich light pink tint, very delicately shaded over the large feathers.

Never having seen a fowl of this colour before, I stopped to look at it, with some suspicion, prepared however to secure it if it really was anything unusual. A Chinaman passing at the time characteristically assumed ownership of the bird, and wished to sell it to me, though I do not believe it belonged to him any more than it did to me. Looking more closely at it, however, I perceived it was a white chicken which had been dipped in aniline dye, and no better nor worse than its neighbours. At Singapore I observed white rabbits dyed with the same pink colour.

The only opportunity I had of searching for marine animals on the Sarawak coast was on a small island called Pulo Marundum, south of Santubon, where I landed on October 7th, and spent some little time at low water. The island is a long, low coral reef, covered with mangroves, among which flew little sun-birds (Cinnyridæ). Small pools and little shallow lagoons contained numerous fish, which the Malays easily caught with their hands; but the perforated rocks were all fixed, and there were very few loose stones which I could take up and examine. A few interesting animals, however, rewarded my search. Among them were those remarkable specimens of Doris exanthemata to which I have already alluded, as equalling in size the largest known individuals of the Nudibranchiate family. Two delicate Planariæ, most difficult to dislodge from the rough coral, occurred here only. One of these was most delicately pencilled all over with little circlets of light brown colour; the other, a large rich velvety brown species, seems to approach the P. zeylonica of Kelaart. Sponges, and botrylliform Tunicates, like clusters of little coloured stars, abounded. Red, tuberculated crabs, of the genus Calappa, ran about, readily taking refuge in the honeycombs of the coral; but, if caught, shutting themselves up as if in a box, from the curious compact manner in which their legs all fitted under the vault of their carapace. Large rough-rayed starfishes (Ophiocomas) also lurked in holes; and I remarked as a fact worth noticing that, fragile as they are, I could, by tugging at one arm or ray, pull one of these *brittle-stars* by main force out from a hole scarcely large enough to admit the passage of its body, and yet without breaking it; whereas, two minutes after, holding it suspended by one arm, it broke short off, evidently amputated by the will of the animal. And, lastly, I found myself the possessor of two new and very beautiful species of Phyllidia, animals closely allied to the Doris, but having the gills arranged in a lamellar form along the sides of the body, instead of on the back, as in the Nudibranchs. One of these was richly tuberculated with shades of green upon a jet-black ground; in the other, the tubercles were of a more simple character—rose-pink in some examples, in others (evidently, however, the same species) of a pale emerald green.

The Pangah, or Head-house, Bombok.

CHAPTER XIV.

THE SARAWAK RIVER.

Eclipse of the Moon—Boats and Rowers—First Halt—Reach the Rapids—The Datu and Chief Hadji—Diamond-washing—Gold—The Last Rapid—Dress of the Dyaks—The Council—Scenery of the River—Mode of Producing Fire—Journey Continued—Incidents—Change Prahu for Canoes—Return Down Rapids in the Dark—Bivouac—Malay Boat-songs—Limestone Cavern—Berlidah—Ascent of Peninjau—Dyak Village of Serambo—Rajah's Summer Residence—Bombok—Return to Sarawak.

A SHORT stay at Sarawak gave me a favourable opportunity of penetrating a little into the interior, and afforded me some interesting glimpses of Dyak life. The Tuan Muda being just about to depute a government officer, Mr. Alfred Houghton, up the river, for the purpose of settling some questions relating to diamond-washing operations, as well as for hearing and deciding some causes among the Malay and

Dyak population, kindly proposed that I should accompany him, a proposal with which Mr. Houghton most courteously complied.

We left Sarawak at midnight on the 24th September in two long boats, or *prahus*, of the kind usual upon the river, propelled by short and broad paddles at either end, while the central part was strewn with mats, upon which the single occupant could recline under the protecting shelter of an *attap* covering of dried palm-leaves, which could be extended or shortened at pleasure. It was the night of the total eclipse of the moon; and during the whole evening a monotonous noise of gongs and tom-toms, accompanied by shoutings, had been going on, becoming more and more uproarious as the eclipse approached totality. In this the Chinese population were assisted by the Malays, with whom an excuse for beating the tom-tom is always eagerly seized; and the mysterious gloom of the unclouded sky was indeed sufficient to attract the attention and arouse the superstitious fears of the untutored population. As the earth's shadow began to show signs of clearing off, the beatings of gongs and tom-toms became however less and less vehement; and a little before midnight the moon shone out in unclouded serenity, and Chinese and Malays alike retired from their task with easy consciences and renewed confidence in the stability of moonshine.

The peculiar and almost mysterious gloom which had pervaded the air all the evening had certainly a depressing effect; and with the voyage in prospect, I was no less glad than the natives to see the moon appear once more, although the eclipse was succeeded by a sharp thunderstorm, in the midst of which we started in our two boats. Owing to the rapids which abound in the higher part of the river, it was necessary that the boats should be small and light, and the crews numerous and skilful. My friend was provided with eight men; I, owing to the short notice, only succeeded in getting five, but these were strong fellows, who maintained their ground creditably against double the number of slighter lads. Their nationality was curiously varied—one was a Javanese, a second from Macassar, a third a veritable Papuan, with the luxuriant hair of the New Guinea race, the remaining two were Bornean Malays. My companion had among his crew two Loo-choo lads, taken as young boys from some conquered junks after one of Mr. Brooke's great piratical engagements, and thus saved from destruction. They had been brought up to honest service, and were very much attached to their master. One of these lads, Sallee, attached himself to me, and a more active and willing pair than this and my Papuan I could not have wished for as attendants.

Once fairly in the stream, all our rowers, who sat with their faces to the front, vigorously plied their paddles, and our boats sped rapidly through the water, and I stretched myself out on my mat for my night's repose; but no sooner were we in motion than the men all began to sing a wild air, keeping time with the strokes of their paddles. This was amusing enough at first, and not without a spice of savagery; but as time passed on and they showed no signs of fatigue in their lungs, it became rather tiresome, for it appeared that they could not paddle without singing, and as there was no reason to complain of the pace at which the boat was going, I was fain to put up with the musical accompaniment, but sleep was banished for the night.

At dawn I was aroused from a slight doze by a pattering rain, and a confused sound of voices, and peeping out from my *attap* covering, I found we were in the midst of a fleet of boats similar to our own, all collected at a common halting-place. Two or three wood-fires were making spluttering attempts to burn under a steady down-pour, and around them were groups of Malay boatmen busily moving about, or cooking little pannikins of rice for their morning meal. These being despatched the boats went on their ways, some up, some down the river; and in half an hour we were alone. After our own boatmen had cooked their rice, which they had well earned by six hours' continuous paddling, we proceeded on our way, the weather clearing as we advanced.

The scenery of the river gradually increased in interest, the low jungle-covered banks assuming a more rugged and elevated character, and limestone hills here and there picturesquely peeping from above the foliage, their steep faces set off in a frame of bamboos and other trees which grew upon the river side. The stream, too, now began rapidly to increase in swiftness, and this was the signal for our boatmen to step ashore for a few minutes in order to cut some stout bamboo poles, the object of which was presently apparent. The paddles soon became of little use in the swift-flowing, shallow river, where the boat had to be skilfully piloted amid projecting rocks, and they were therefore now laid aside, and our boatmen at once gave proof that they were equal adepts in the art of poling. Their management of the boats was perfect, and the rapids which we now began to ascend afforded them ample scope, though it was afterwards in descending them that their dexterity was chiefly conspicuous.

As we proceeded we were joined by several other boats, all bound for the same destination as ourselves; for it was known up the river that Mr. Houghton was empowered by the Sarawak government to make some arrangement for the working of the diamond-washings, and many who were interested in the subject joined our procession. Among them was the Datu, or first native chief of the district, and the chief Hadji or Mahommedan priest. The Datu was a pleasant-looking man of about fifty, who has supreme authority in the native courts in petty cases, divorce suits, etc., and was described to me as a man of considerable intelligence and sagacity. He had with him his son, a slight lad of nine or ten years old, and like his father of intelligent look, and gentlemanly manners and appearance. The Hadji was a stout, heavy, and somewhat dirty-looking old gentleman, disposed, however, to be very friendly. Unfortunately I could not talk much with them except through my friend, my knowledge of Malay being unequal to the effort of sustained conversation; but both, particularly the Datu, made themselves agreeable, and as conversational as the circumstances would allow.

In a bend of the river we came upon a number of the diamond-washers at work, and stopped to observe their mode of procedure. Diamonds have long been known to exist in the river-bed, and the search for them has been carried on for a long period. For the most part they are of small size, but of a brilliant water, although large ones have been occasionally met with. The largest Bornean diamond belongs to the Sultan of Matan, and is valued at £269,738, weighing, as uncut, 367 carats. In the sand and gravel of the river-bed, at depths averaging from six to eighteen feet below the surface, and in strata sometimes several feet thick, the diamonds are sought for with varying success by a large number of Malays, who sink shafts at a distance of 20 feet apart in the shallow parts of the river.

They construct huge pyramidal frames of large and strong bamboos, about three yards square at the base, and by means of heavy stones they sink them upon their claims, so that they may not be carried away by the stream, and at the same time shall point out clearly the working-place of each party. Their most important stock-in-trade consists of a number of large round bowls of wood, extremely shallow, and ingeniously cut, as if with a turning-lathe, deepest in the centre, and shelving all round to the rim. Filling this bowl with gravel, etc., from the river-bed, they (standing in the water) hold the bowl, just skimming the surface, and give the contents a rotatory motion, cautiously and skilfully allowing the muddy and lighter sandy particles to flow over with the water, until nothing is left at the bottom of the dish but the larger and heavier sandy and gravelly substances, which are then carefully examined for the coveted diamonds. This work can, however, only be carried on at certain periods of the year, namely, during the dry season, for during the rains the river so swells as to render it utterly impossible to make any attempts; and this will be understood when it is mentioned that at the spot where we encamped, and where was a small settlement of diamond-washers, I was assured that the river at certain periods rose 30 feet higher than it stood at the time of our visit.

Besides diamonds, gold is found in tolerable abundance in Sarawak, much being obtained from the same situation and by the same means as the diamonds, chiefly by the Malays. The Chinese population find much gold in the alluvial soil —a yellow, clayey loam,—and this is the best and most reliable source. Altogether, at least a picul of gold ($133\frac{1}{3}$ lbs.) is procured annually from these sources. Silver, however, has never been found in this district.

At length we arrived at a long rapid, which really deserved the name, and taxed the powers and skill of our boatmen to the utmost. With loud shouts they used their bamboo poles in good earnest. Our numerous paddlers were all on the alert and in a high state of excitement; still they could hardly make any way against the rush of water. Overboard they leaped pell-mell, and although they could scarcely keep their legs, they lugged the boats by main force through the narrow channel. Those that were best manned of course got through first; but my five men, although they could well hold their own against crews of eight or ten in smooth water, were hard bested to get my boat through the critical place. But they were soon reinforced by those in advance, who came back to the rescue, and with excited shouts dragged it up the cataract. Meanwhile the smaller boats with fewer rowers crept laboriously up the side channels in-shore, and soon all were safely through, and at noon we found ourselves all moored by the steep banks of the river, at a little impromptu village of diamond-washers, which was to be our halting-place for the night.

After a pic-nic breakfast and a bath in the river, we felt sufficiently refreshed to enter upon the duties which had brought us to this spot. It was a lively and interesting scene, and was rendered more so by the presence of a family of Dyaks, consisting of a man and woman, and a girl of about 18. The dress of the man was simply a handkerchief upon the head, and a scanty cloth wrapped round the loins. By his side was the universal *parang* or chopping knife, and the *tambuk*, or bag from which he took betel and sirih. The woman wore only a short petticoat, tightly fastened around the lower part of the hips and hanging to the knees. The arms and legs were encircled by spiral coils of brass wire from the shoulder to the elbow, and from the elbow to the wrist, leaving the joint free; and from the knee to the ankle, so closely compressing the limbs that the flesh bulged out in an unsightly manner where the compression ceased. These wires appeared incapable of being removed; indeed, I afterwards saw a woman washing in the river with these incumbrances to cleanliness, which must be very provocative of ulceration also, upon her. A number of thin rattan rings encircled the girl's body, which was otherwise bare, and some rings of rattan and of brass hung around her neck. Her black hair, loosely parted in the middle, hung loosely behind as far as the waist.

The first view of these aborigines gave me a strong impression of the wretchedness of savage life, as they lingered to hold some conversation with the Malays, and cast some curious glances at the white men, from whose gaze the women seemed instinctively to shrink.

Meanwhile, all was prepared for the Council, and the government official having taken his seat, with the Datu, upon one of the diamond-washers' frames by the river-side, and surrounded by the natives who were interested in the matter, opened the question and stated the arrangements which the Sarawak government had determined upon for the fair apportionment and regulation of the various claims. He was listened to with intelligence and attention by the Malays who squatted round, chewing betel, and then the matter was fairly and quietly discussed, questions answered, and details explained to the satisfaction of all parties, after which the meeting broke up. Meanwhile our servants were building a little shed upon the hill-side, and covering it with attaps for our accommodation during the night.

The scenery of our camping-place was exceedingly picturesque. Ascending the steep bank of the river, magnificent limestone crags could be seen peering out through the luxuriant foliage, and calling to recollection the Trossachs in Scotland. The rapid river, having just descended an incline, which had the steepness and swiftness of a cascade, was hurrying on to the long rapid which we had that morning ascended with so much difficulty. A number of Malay boys in canoes were amusing themselves by dragging their boats up the rocks beside the cascade, and then shooting it with the utmost fearlessness and dexterity amid shouts and laughter. Some exquisite butterflies flitted by every now and then, affording occupation to my net, as I rambled on the top of the high bank, upon which a few huts of the diamond-seekers were perched, while the washers were already at their work, standing in the river with their shallow dishes in search of the gems.

A heavy shower of rain having driven us to the shelter of our attaps, we sat and amused ourselves with chatting with the good-natured Malays who accompanied us, and who were ever ready and willing to do us any kind offices. I seized this opportunity of learning the mode of producing fire, which is seldom described, but usually taken for granted as known. My request that they would make fire was answered by one of the Malays selecting from among our firewood a dry stick of hardish wood, about 15 inches long, which he cut with his *parang* into the form of a thickish lath, and having also made a small notch on the narrow edge, stick number one was ready for use. Taking a smaller piece of wood of the same kind, about nine inches long, he pared it into a cylindrical shape, and cut one end straight off. Then placing the long stick on the ground with the flat side uppermost, and setting his feet firmly upon the two ends, he put a piece of paper under the notch, and taking the small stick between both hands, as he squatted before it, adjusted the flat end to the smoothed surface of the larger stick immediately adjacent to the notch. He then rotated the small stick rapidly between his hands, pressing it down upon the larger one, until by degrees a round hole was formed, and a ligneous powder was produced, which fell down the notch and formed a little heap upon the paper. After having thus rubbed for about two minutes, the powder began to smoke, and then turning black as the increasing heat charred it, suddenly became red-hot, and the tinder thus formed only required a puff of breath at this critical moment to ignite the paper beneath. The exertion required was considerable, but of short duration.

On the following morning the report of some coal which had been discovered by Dyaks higher up the river induced us to make arrangements to visit the spot; and accordingly, immediately after breakfast, accompanied by the Datu and the Hadji, we set out with a single boat for an excursion higher up the rapids. The first of these was immediately above our halting-place, and the steepest we met with, forming indeed a complete cascade between two projecting rocks. The boat having been dragged round, we proceeded through some highly picturesque scenery, the river becoming more shallow and the rapids more frequent as we advanced. Many beautiful precipitous limestone rocks towered around us, of which those called Gunong Gigi and Retti were perhaps the most striking. The banks were at one time steep and rocky, at another low and alluvial, and upon them were numerous gardens cultivated by the Dyaks, and containing bananas, cocoa-nuts, and other trees. The little mouse-coloured swallow (Hirundo esculenta), which forms the *birds'-nests* of commerce, frequently found in the numerous limestone caverns with which the country abounds, accompanied us all day, skimming like the English sand-martin over the river surface. Another little bird, having a note not unlike that of the yellow-hammer, was pointed out to me by the Malays as the alligator-bird, about which they had a legend to the effect that the alligators of the river were constantly demanding of it payment of a debt long due to them from its ancestors, to which the bird is supposed to reply, "I have nothing to give you except the feathers of my tail, and those you may have if you can get them,"—a legend which seems intended to place their most dreaded enemy in a ridiculous light. For these terrible monsters ascend the rivers even above the rapids, and, if well-informed residents can be trusted, even sharks have been known to do the same. The Malays remarked of another small bird that it was the diamond-bird, averring that wherever it was seen diamonds were certainly to be found near at hand.

At a bend of the river we suddenly met a canoe, paddled by four little Dyak girls, who seemed half frightened at meeting us. I was much struck with their upright graceful figures as they sat at their paddles; for, like other savage or semi-savage people, the children are often pretty, though the exposure which they undergo, and the labour which they necessarily have to perform, soon turn their infantine beauty into the harsh and unredeemed ugliness of coarse adolescence.

The increasing shallowness of the river rendered it necessary soon after this that we should change our boats for two small canoes, in each of which two of us sat, with one of our boatmen at the stern, while a Dyak wielded a pole in the fore part of each, and thus we performed the remainder of the journey. Monkeys were occasionally visible in the trees as we passed, but all of small species; for the large red orangs do not frequent this river, though they are common in the neighbouring river, Sadong. Hornbills screamed as they flew from tree to tree, and numerous handsome butterflies flew across our path; but the most striking insect was a dragon-fly, with wings of a metallic golden-green, which glittered gorgeously in the sunlight, and was by no means uncommon.* Little, scarcely visible, sandflies, also, were rather too numerous, and inflicted their bites on any exposed parts of the body like sharp needle-pricks.

We halted at a sandspit for a bath, and to allow the Datu (who was a zealous Mussulman) and the Hadji opportunity to perform their devotions, which they did after their ablutions, with their faces towards Mecca. Above us on either side of the river we could hear voices, which we found to proceed from Dyak villages almost concealed among the trees upon the high banks. From the houses bamboo platforms, supported upon poles, ran out, upon which we could discern some dusky forms watching us from their shady retreats; but no particular surprise or excitement was caused by the visit of white people, to whom they are all more or less accustomed.

At length we arrived at our destination, and landed just as a heavy tropical shower came on. Penetrating the jungle which bordered the river, under the guidance of a Dyak, we very soon arrived at a spot where a seam of indifferent coal cropped out from the thick layer of vegetable mould, and

---

* On subsequently looking over the collection of insects made by a gentleman in the island of Hong Kong, I was surprised to observe among them a species similar if not identical with that above described, which he told me he had frequently met with in Happy Valley.

having secured specimens, at once prepared to return. After two hours' swift paddling down the stream, the light began to fail; and as the moon rose late, we were for two more hours shooting the rapids in almost total darkness, for the trees overshadowed the stream so as to cut off the feeble light of the stars. It was here that the skill of our Malay and Dyak boatmen was tested. So completely were they masters of the canoe, and so thoroughly acquainted with every bend and fall of the river, that we proceeded with perfect safety, although in the event of our missing the channel, or striking against a stone, our slight skiffs must in some places inevitably have been upset. There was nothing for it, however, but to gather ourselves well up in the middle of the canoe, and sit still, with full reliance upon their powers. I was not sorry, however, when we at length reached, in total darkness, the spot where our large boat had been left, soon after which the moon rose, and the rest of the way was plain sailing. Late at night we arrived at the last and steepest rapid, which really required skill even by moonlight. Holding on by the gunwale I steadied myself for the descent, and with a shout from the rowers, we shot down, scarcely shipping a drop of water, and were once more at our camping place. The bustle of our arrival soon brought our servants with lights, for, not expecting us, they had already retired under the attaps; but a fire was quickly kindled, and supper prepared, one ingredient of which was the young fronds of a fern, a species of Marattia, which, when boiled, makes an excellent and palatable dish. Our mats were spread, not upon the ground, as it fortunately happened, though in close proximity to it, for having stepped off, barefooted, I was nipped and bitten in such a manner that I was glad to beat a hasty retreat, and on holding the lantern to the ground it was seen to be swarming with large ants. I had great fear of a visit from them during the night; but although only 18 inches below my couch, they fortunately abstained from climbing up the posts, and left us unmolested.

The following morning we quitted this spot, and formed a procession of boats down the river. The boatmen, as usual, enlivened the way with their songs, some of which were wild and musical. They all joined in the chorus; and one of them, of which they appeared particularly fond, had a refrain which ran as follows, the *staccatos* being strokes of the oar:—

Keeping time with their paddles, the song was cheerful and inspiriting, and seemed to help them along. There was no end to their good-humour and spirits, and they delighted in a sharp race on the smooth reaches of the river. At the slightest challenge they would ply their paddles at a prodigious rate, skimming along with shouts and cries of encouragement, and at a speed which would have done credit to "dark blue."

Soon after mid-day we halted, that we might land and inspect a fine limestone cavern, of whose existence we had been informed by some friendly Dyaks. Under the guidance of one of them we proceeded along a well-trodden path through the wood, and in about ten minutes reached the skirts of a Dyak village. A deep, narrow gorge between limestone rocks had at the bottom a pool of clear water, being all that remained at this season of a torrent which had evidently, in course of time, worn away the chasm; and in this pool a Dyak woman and some children were bathing, and looked up in astonishment at the large party of intruders skirting the edge of the ravine. Following this it led us directly into the cavern, the entrance of which was grandly arched, and somewhat recalled the entrance to the Peak in Derbyshire. The interior was very spacious and irregular : the left-hand side was hollowed out into intricate chambers, with irregular natural steps leading from one to another, and loopholes connecting them. The Dyaks having provided us with flambeaux of bamboos, we entered and explored these chambers, disturbing a number of large bats which had taken refuge there for the day. All that side of the cave was heaped up with an irregular and deep deposit of an alluvial character, while on the right-hand side a deep channel was cut in the rock by the stream, which entered at the other end of the cave, and which, having once been large enough to excavate the whole cavern, had now dwindled to a rivulet flowing through this narrow, rocky channel, which was also continuous with the deep ravine first alluded to. From the roof of the cave, in many places, depended enormous stalactites many feet long, and of a diameter sometimes exceeding that of a man's body. The cause of this appearance was not difficult to be explained, for above the cavern trees were growing, the roots of which had penetrated the soil above and sent down fibres through the roof; the water percolating along their course had at once encouraged their growth and elongation, and carried down calcareous particles which encrusted them as they grew, so that each enormous stalactite was a network of rootlets entangling between them, and encrusted with, masses of lime.

Of course I was in no position to undertake anything like excavation in this cave, much as I should have liked to learn whether anything was buried in the mass of alluvium deposited; its extent and position would render it a costly and difficult enterprise. Many similar caves are known to exist in the limestone hills upon the Sarawak river, but this particular cavern was unknown even to the Datu, who had lived all his days in the neighbourhood.

On reaching the west branch of the river we parted company with the Datu, and pursued our way towards Berlidah, a bungalow belonging to the Rajah, and beautifully situated at the foot of Mount Peninjau. As we were approaching it we were overtaken by a most tremendous tropical storm, such as often comes on early in the afternoon at this season. The most vivid lightning was accompanied by the heaviest rain I ever beheld, in the face of which our unprotected rowers sat plying their paddles, while the great drops pelted mercilessly upon their bare skins. As for ourselves, drawing the attaps close over us, and covering ourselves with coats and blankets, we contrived to keep tolerably dry through this deluge.

Having arrived at Berlidah, as though not wet enough, our men plunged at once into the water, while we disembarked and soon were established in the bungalow. It was too wet to do more than sit in the verandah, and see the Dyaks returning in their canoes to the foot of Mount Peninjau, at the top of which more than one of their villages was situated, or to listen to the Wou-wou (Hylobates), a small and pretty species of Gibbon, so called from the peculiar noise it makes, like that produced by pouring water from a narrow-mouthed bottle; or to watch the fire-flies when the darkness of night had supervened, flashing intermittently at the edge of the dense jungle which covered the mountain.

Next morning the mists rose up, as I have seen them do from an English valley, giving promise of more settled weather. So as soon as I had despatched my Malays in search of land-shells and butterflies, taking my Papuan as a guide, I ascended Peninjau, in order to visit the Dyak village of Serambo. In the early part of the morning the Dyaks, having descended the hill, passed us in some numbers in their canoes, going on their daily avocations in the valley, and some few we met as we mounted the hill. The road was very steep, like a gully in a Welsh vale; the stones and rocks extremely rugged, and putting our climbing powers to the test. Every here and there, when they were more perpendicular than usual, stumps of trees, with notches roughly cut in them to serve as steps, and called by the Malays *batangs*, rendered the way more practicable. At length, after a climb of two hours, we reached the shoulder of the hill, on which was situated the village, consisting of scattered dwellings almost buried among palms and bananas. The houses are in groups or sets, built of wood, bamboos, and nibongs, and are raised from the ground on posts, the houses of each set communicating with one another by doors between them; and in front of them is a long, elevated platform of bamboos, which is reached by means of *batangs*. Such houses are built wherever a footing can be found for them, without any regard to arrangement; and such is the rugged and overgrown character of the locality, that they are placed at all levels, and are partially concealed from one another by the trees and rocks, so that the extent of the village is not readily distinguishable. In order to facilitate the passage from one house to another, stepping-stones are placed among the long grass to guide the feet, which would otherwise be betrayed into stony crevices, and similar awkward traps; and some of the larger and deeper holes have to be crossed by rude bridges, consisting of a number of bamboos stuck into the ground in the form of St. Andrew's crosses, a narrow plank being laid upon the angle between them. Water flows freely from the upper part of the hill, and is conducted in bamboo pipes to a convenient spot, where there is, consequently, a constant supply.

About these houses were a number of timid, dusky, half-naked women and children, the men, for the most part, having gone down to the plains. They seemed rather frightened at the invasion of their solitude by a white stranger; but, thanks to the government of Rajah Brooke, the time has long gone by since they were liable to be robbed and carried away into captivity when surprised by neighbouring tribes in this unprotected condition. I entered some of the houses, and found some women within, either nursing or with children playing around them, and by means of little presents easily succeeded in calming their apprehensions. The interior was tolerably clean, each having a stove and cooking utensils, and a mat spread in the corner, which inventory, however, appeared to embrace the whole of their household furniture.

In this village was one house which arrested attention from its peculiar form. It was circular, built upon piles, and had a conical and pointed roof, and several windows or rather holes to admit light, protected by shutters. This was the *Pangah*, or head-house, an institution in every Dyak village, and in it used to be deposited the heads or the skulls of their enemies, or, more strictly speaking, of the strangers captured in their head-hunting expeditions. But such head-taking practices are now entirely abolished in the territory of Sarawak. The head-house is spacious and roomy; and when a stranger asks for hospitality among these people he is always lodged in one of these places, as the most eligible house in the village.

On the summit of the hill above Serambo is a pleasantly-situated bungalow, belonging also to the Rajah, and occasionally resorted to as a sanatorium. The approach to it from the village is very steep, and is formed by regular steps cut out of the hill-side. From this summit, and from the verandah round the house, a splendid and extensive view of the country was afforded, a view singularly characteristic of a great equatorial region, sparsely populated and little cultivated. Far and near, wherever the eye roved, the jungle extended dense and unvaried, except by the undulations of the land, and here and there by some small grassy spots, rendered striking by the absence of trees; but these spots bore the same relation to the jungle that clumps of trees would do to an English meadow. Through this endless forest meandered the Sarawak river; but no villages could anywhere be discerned, the Dyak houses being everywhere concealed from view. The sole trace of civilisation in the landscape was the bungalow at Berlidah, which could be seen imbosomed in trees by the river's side. The salient forms and craggy sides of the various limestone hills of Singhi, Matang, &c., gave some variety to the scene; while coming from the dense wood behind where I stood, I could hear the laughing and chattering voices of Dyak children in the village of Bombok, a second hill-settlement, which I visited next day. From this village two stout Dyak girls, whose scanty clothing was confined to a short petticoat, appeared upon the scene while I was gazing at the prospect, and without any timidity accepted some of the refreshments, in the form of young cocoa-nuts and oranges, which had been dispensed to me by the Rajah's housekeepers.

On the way down I met several family parties of Dyaks toiling up to their homes, carrying on their heads heavy bundles of grass, and other things which they had procured from the lower world. I could not help feeling that, although there had been a period when it was safer to dwell in the mountains, it must be fearfully inconvenient and trying to perform daily such a laborious journey.

On the following day I ascended Peninjau from the opposite side by a new route, in order to visit the village of Bombok. This ascent was even steeper than that to Serambo, the stones were higher, the rocks more rugged, and the *batangs* longer and more perpendicular; so that in the absence of anything to hold on by, it was not easy to mount them, although I had no truss of hay upon my head. The operation was not unlike walking upon a tight rope, especially as the heavy rain, which had drenched me as I returned the previous day, had rendered the notches slippery, a circumstance keenly felt in the descent. It was while balancing myself cautiously upon one of these precarious *batangs* that I was suddenly startled by a loud rushing noise overhead, which seemed rapidly approaching, and gave me a momentary apprehension that a tree was falling, or that an avalanche of stones was rolling down from above. Looking up in some alarm, I discovered that the sound was produced by a large hornbill (Buceros), which was performing a drumming noise with his great wings, precisely similar in character to the drumming of the snipe. The hornbill, hovering in the air, vibrated the quill-feathers of the wing with violence and rapidity, thus producing the rushing noise which had so startled me.

Bombok appeared to be a more populous village than that on the opposite side of the hill. We met numerous parties descending to their daily work—women and girls as well as men; the former, on catching sight of us, sometimes threw down their bundles and ran with a cry into the jungle. So similar, however, was it in character and in details to Serambo that I could scarcely believe that it was not the same village which I had visited yesterday. The headhouse occupied its usual conspicuous position, and was of the same form as before. The interior was a spacious and dusty apartment, in which were a few skulls hanging up here and there, their mouths plastered up with mud, and cowries occupying the place of eyes. Here were but few, however, the place having fallen into disuse from the discontinuance of the barbarous custom of headhunting.

While I was inspecting the head-house every window or outlook in the village was filled with the dusky faces of women and children; and having a few small mirrors in my pocket, I exhibited them with the announcement that I would give one to any little girl who would come and take it from my hand. For a considerable time even this bribe would not induce them to come so near; but it succeeded at last, and when one had come forward and returned with the prize, I had plenty of candidates, and my supply was soon exhausted. They then became friendly enough; and as I left the village to descend once more I was followed by a number of children of both sexes, who timidly, and with looks and whisperings of curiosity, escorted me to the outskirts of the village.

On arriving at Berlidah I found my boat ready for the journey back to Sarawak, for my time had expired; and after some hours' paddling down the stream, Government House appeared in sight, and we reached the ship just in time to get shelter from one of those heavy tropical storms which at this season so often pass over in the course of the afternoon.

CHAPTER XV.

SINGAPORE.

Variety of life in Singapore—The Malays—Their Villages—The Klings—Kling Women — Their Occupation — Religious Ceremonies—Mosque—The Chinese—The Bugis — Residences— Native Streets — Tigers, not numerous—Fire-Flies—Botanic Gardens—Sensitive Plants—Kling Bird-Catchers—Climate of Singapore—Productions of the Sea Shore—Sharks.

THERE are few places which present such variety of scene, and ever-changing novelty of Eastern life as Singapore. Situated almost on the equator, a great central emporium of trade, and a meeting place between India and Europe on the one hand, and China and the far distant regions of the Malay Archipelago on the other, its geographical position renders it one of the most interesting places in the world to the observant student of nature. Here we may see tropical vegetation in all its beauty and perfection; and here, too, we may meet representatives of various races from the east and from the west, attracted by the same commercial magnet,—Europeans and Asiatics all alike bringing with them their manners and customs, their religions, their costumes, unchanged—a picturesque combination such as scarcely any other place can afford.

The Malays are the real indigenous sons of the soil, and contribute not a little to the general effect. They are not a handsome race. Their mahogany-coloured faces and high cheek-bones are usually accompanied by a remarkably shapeless and ugly mouth, which is rendered even worse by the detestable habit, common to both sexes, of chewing betel-nut, which reddens the teeth, lips, gums, and saliva of an uniform blood-colour, and has a most unsightly appearance. They usually affect bright colours in their costume—the men wearing a *baju* or jacket of thin material, more or less variegated, and trowsers, *sluar*, of a similar character, while the head is enveloped in a *saputangan* or coloured handkerchief. With the women the *sarong* (literally a sheath) plays an important part. In less frequented places it is the sole article of dress, and consists of a wide skirt or sack, of equal size above and below, fastened just beneath the breasts and reaching to the ankles, the shoulders and arms only being left bare. In young girls the little sarong is commonly of a yellow colour, and indistinguishable at a distance from flesh-tint; it is simply fastened round the waist, and is the only garment, and a very graceful one, exhibiting the contour of the figure, especially when, as in some cases, it is ornamented with a quasi-classic pattern, strongly reminding one of the antique female dresses which we meet with in Hope's "Costume of the Ancients."

The Malays are very lazy, and averse to any, especially continuous, labour; and scarcely anything can induce them to undertake active employment. The women have universally a listless, shuffling gait, and languid appearance, which is very characteristic, and not improved by the use of slippers, which simply hang upon their toes without any fastening.

A Malay village is a common but picturesque and interesting feature of the neighbourhood of Singapore. Its situation and surroundings cannot fail to strike a stranger, and to be a matter of interest to the observant. These villages are always built upon platforms raised on wooden piles, either on the margin of the sea, or more commonly of rivers and creeks, or over mangrove swamps; and the water in all cases, at some period of the day, flows underneath their dwellings. It might be supposed that such damp situations would be unhealthy in the extreme; but such are always chosen by these people, who build clusters of wooden houses called *kampongs*, which are either approached in boats, or by a pathway of earth, raised above the level reached by the rising water. A background of cocoa-nuts and bananas often adds beauty to the scene, and the dusky forms and faces moving about among the rickety dwellings, on the platforms, or paddling about in canoes, give it a curious semi-savage aspect. All the litter and dirt of the establishment is simply swept through the open cracks in the floor, and without further trouble is thus washed away by the tide; an arrangement which well suits the character of the indolent Malays, who lazily lie upon their backs on the platform, while the little naked children make dirt-pies, or paddle about in minute cock-boats; the women, meanwhile, listlessly shambling about on their household duties. Such a village has a wild and uncivilised appearance, and yet there are not unfrequently curious and incongruous signs of luxury visible. Outside some of these houses may be seen vases of flowers, well cared-for; and I have noticed here and there cushions laid out apparently for airing, whose ends were daintily embroidered.

Sauntering by myself one day through such a village on Pulo Brani, I was not a little surprised to hear sounds proceeding from the interior of one of the houses, of an infantine voice, which said p-i-g, pig, d-o-g, dog, &c. It was evidently a child in the throes of the first spelling-book; but in what an uncivilised spot! Wishing to understand the cause of this phenomenon, I walked into the chamber, where I found a respectable Chinese lad of 15 or 16, who had been educated at the High school of Singapore, and spoke English very well, and was now teaching his little brother, 7 or 8 years old, the mysteries of the barbarian language.

Owing to their proximity to the water, the children are veritably amphibious in their habits. They hollow out logs of absurdly small dimensions, which do duty as canoes, and these they propel either with an impromptu paddle, or even with their hands; and they will dive like ducks, regaining their frail craft with astonishing skill. Such canoes often come up and surround the ships in New Harbour, their occupants diving for small coins thrown into the water, which they never fail to secure.

The foreign (Eastern) residents in Singapore mainly consist of two rival races, widely different in dress, habits, and religion—viz., Klings, from the Coromandel coast of India; and Chinese. I say rival races, because they are both active and industrious, and compete with one another in the chief industrial occupations by which a livelihood can be obtained. The Klings are, indeed, the only people who can contest the field with the Chinese, and they do so notwithstanding many disadvantages. They are intensely black—not the shining black of a negro, but a dull sooty colour, from which their eyes gleam out with great expression, half savage, half intelligent. They are remarkably well-built men, tall, slender, clean-limbed, and graceful, and their faces are often positively handsome; the features small, and finely chiselled; nose aquiline; mouth small, and teeth white; a highly intelligent cast of countenance, which, translated into a white skin, would be considered elegant and fascinating. Their hair, which is black, straight, and glossy, is shaved off the forehead, giving them a commanding look; but is allowed to grow long behind, and usually gathered up into a knot. Their dress, when they indulge in any, is highly picturesque; but they not unfrequently wear nothing beyond a linen cloth around the loins. But when they do dress it is either in dazzling white, or white combined with some bright colour, particularly scarlet; a capacious white mantle, twisted in a complicated manner around the whole body, and surmounted with a white or a scarlet turban. They usually go barefoot, but sometimes wear wooden sandals with a button inserted between the great and second toe; or large, purple leather slippers, trodden down behind and turning up in front in a long curved point.

The Kling women are dark beauties, finely made, and dressed in flowing robes, which conceal the whole figure down to the feet, but leave the arms bare to the shoulder. Their dress sits on them gracefully, and their ornaments give them an air of barbaric splendour. Armlets of gold are worn above the elbow, and bracelets of gold upon their arms; golden rings encircle their ankles, and several finger-rings glitter on their hands; heavy ear-rings hang pendent from their ears, and one side of the nostril is pierced to give passage to a gold nose-ring, more or less chased in front. These ornaments are not unfrequently all worn by one woman, and it appears to be a common practice to invest their money in these trinkets, so that a Kling woman carries a small fortune upon her person. But cases are not wanting in which this ostentatious display of wealth has excited the cupidity of miscreants who have murdered the woman for the sake of what she carries, and have ruthlessly torn the rings from her ears and nose, and wrested the armlets and anklets from off her limbs; and the wonder only seems to be that this does not more frequently happen. The little black Kling children of both sexes dispense with all clothing more commonly than the Malays.

The Klings are universally the hack-carriage (gharry) drivers, and private grooms (syces), and they also monopolise the washing of the clothes of the dressed community. The Singapore Dhobies, as they are termed, who are nearly all men, certainly produce whiteness in linen, but sadly at the expense of material. Standing in a stream beside a large flat stone, they rinse and soap the clothes, and then, instead of rubbing them in the fashion of Western laundresses, they dash them over their head repeatedly with great force, and with a loud sough, upon the stone, which, however rough it may be at first, is soon worn smooth by the contact of the unfortunate linen.

But besides this class, there are Klings who amass money as tradesmen and merchants, and become rich. I visited their silk shops, in the busy part of the town, kept by men who from their scanty and poor attire were scarcely distinguishable from coolies, but nevertheless had a balance at their bankers', and were in the habit of buying goods to the amount of thousands of dollars at a time from the merchants with whom they deal. One was pointed out to me who had just purchased for 5500 dollars a small cocoanut plantation, which he was anxious to possess, simply because, since it adjoined his own property, he did not wish another to hold it.

The wild, barbaric habits of these people were, however, best exhibited in their religious ceremonies. Having observed a brilliant display of lights on the shore for several nights, and great beating of tom-toms in the same vicinity, I threaded my way through the crowded Chinese streets, and guided by the glare, found myself in front of a Kling mosque of the usual tawdry description, with miniature windows, and two small minarets, the whole front whitened, and hung with numerous lamps, consisting of small glass tumblers of oil. Crowds of Klings were going in and out, and loitering before the mosque; and a great many Chinese were gathered as spectators. On the opposite side of the street was a picturesque group of Klings seated cross-legged in a semicircle, each alternate one having in his hand a large tambourine, and the central seat being occupied by one who appeared to be the president or chief priest, in front of whom a fire was burning in a brazier, which was fed with combustible matter from time to time, blazing up and throwing a picturesque glare upon the swarthy figures seated around, the standing crowd of Klings behind them, and the outside mob of Chinese, who were spectators.

Presently a man came forward with two awl-shaped instruments, which having first smeared with ashes from the brazier, he deferentially presented to the chief priest, who taking them in both hands, closed his eyes, and muttering some words over them, returned them. The man then began to dance, while the tambourines were struck in a measured manner, the strikers swaying themselves backward and forward, and becoming gradually more and more excited, until at length the perspiration poured down their bodies. Meantime the devotee in the centre danced on with frantic gestures, and flourishing his two weapons in his hand, he thrust them (or pretended to thrust them) into his belly (his only costume was a cloth round the loins);

then into his arms, making a show of wiping off the blood after each thrust—but I was pretty sure at the time that it was entirely sham, and that he did not wound himself in the least.

Then taking a sword and ostentatiously feeling its edge, he presented it to the priest in like manner, and the dance recommenced. Flourishing it about over his head, he placed the edge (which appeared to me to be very blunt), against his bare shoulder, and with a small wooden mallet which he swung round with the other hand he struck the back of the sword. The tambourine-players kept triple time, accompanying the performance with a loud, wild, and monotonous song, but I do not believe the man wounded himself, although he repeated the act several times. The exhibition then closed, during which the Klings standing around showed us great attention, making the Chinese move aside, and inviting us to come forward, where we could better see the proceedings. We afterwards crossed over to the mosque, which we were invited to enter, and having complied with their custom of removing our shoes and exchanging them for slippers, we did so. Here also the Klings were officiously polite. The interior of the mosque was small and narrow, and, like the exterior, brilliantly illuminated. The whitened walls were decorated with tinsel and red and green paint, in a very tawdry manner, but the effect was not disagreeable, though somewhat theatrical. We were not allowed to enter the inner chamber at the back; and presently retired—the attention of the Klings being continued to the last.

The Chinese inhabitants of Singapore have all the characteristics of Chinese elsewhere. They are the busy working population, who hive together like bees, performing the artizan duties necessitated by a large community, and other essential operations, of a nature which the more scrupulous Klings would disdain to lend their hands to. Street after-street is crowded with these active and energetic Celestials, who here, as in China, have the spirit of centralisation and co-operation. One street contains none but blacksmiths toiling over their forges—another resounds with the saws and chips of a crowd of carpenters—a third with the dull blows of the stone-masons' mallets, &c. Each devotes itself to its own craft, and admits no interlopers, while all the inhabitants are busily engaged in effecting one common end,—and that is, to hoard up enough money to enable them to *cease work*, and return to their native district in China, there to pass the remainder of their days in a comfortable independence among their own people, and especially amidst their own family.

Besides these there are, in certain parts of the town, considerable numbers of Bugis people from Celebes, &c.,—tall and strong, wild-looking fellows, wearing a pair of short drawers and a rough cloth flung over the left shoulder like a plaid. They belong chiefly to the *prahus* or crafts which lie at the east side of the harbour, and may be generally seen loitering about at this end of the town out of mere curiosity, and strongly reminded me of the Lancashire operatives wandering about Liverpool, open-mouthed, on a holiday. Their vessels are ugly-looking, two-masted craft, with black and white longitudinal stripes, which trade with Singapore from all the islands of the Archipelago. Formerly there were many more of them, for while Singapore was a free port, all the Dutch ports required dues and levied imposts. The Dutch, however, learning wisdom, though late, freed their ports; and since then they have diverted great numbers of the Bugis prahus from the Singapore market, to which those which still repair thither bring gold-dust, tortoise-shells, ambergris, pearls, birds'-nests, turtles' eggs, sharks' fins, trepang, mother-of-pearl shells, and other curiosities of Eastern commerce.

The bungalows or residences of the European population are generally set in the midst of gardens or small plantations —pleasant habitations, surrounded with Betels and Bananas. They have, for the most part, an imposing look, for people in the East can not, or will not, live in small houses; and the open verandahs which nearly always run round the outside, make them look larger than they really are. For the most part such residences are exclusively occupied by Europeans, neither Malays nor Klings live in such expensive style; but certain wealthy Chinese exceptionally make a great show, and entertain in a style of costliness and refinement not surpassed by the most opulent English. Of these Mr. Whampoa, who contracts with the Government for the supply of stores and provisions to Her Majesty's ships, is *facile princeps*. He is well known for his choice and elegant entertainments, and his mansion is surrounded with extensive gardens laid out in true Chinese style. They have a light and graceful aspect, being devoid of any heavy masses of shrubbery as in English gardens. Rows of Betels and feathery Casuarinas spring up here and there, and the paths run like mazes among the beds, which contain many choice flowers. Here and there on the terraces we come upon quaint devices of trees trained over a framework to represent a bird, a fish, or a quadruped. ·Pleasant ponds enliven the scene, containing water-lilies and Nelumbiums; and one of them, especially large, is devoted to the Victoria regia, the queen of water plants, where it thrives well, having flowers and buds, and leaves seven feet in diameter.

The native streets of Singapore consist of two-storied houses, with an arcade running along in front, narrow and more or less dilapidated, and blocked up with merchandise. In these the Chinese and Klings live side by side, the houses of the latter being usually less looked after and more ruinous. In their shops are often exposed European articles of the commonest kind and poorest quality, which, however, find a ready sale. In the outskirts of the town and less built-up streets the scene is interesting and picturesque. Tall and graceful Betel Palms, spreading Bananas and Cocoa-nuts, dense and feathery Bamboos and Casuarinas, bushy Screw-pines (Pandanus), handsome-leaved Bread-fruit and Jack trees (Artocarpus), Durians and Mangos, and other fruit-trees constitute the leading features of the vegetation. Mangroves twine their long, rib-like roots in marshy spots, and Melastomas, Heliotropes, sensitive plants, and luxuriant ferns characterise the drier spots. Amid these scenes the variously-costumed people are a never-ending study: the Chinese, generally ugly and dirty, with scanty clothing usually of a dirty colour, and nearly always with a load upon their shoulders,—the women, and often little girls, carrying an infant astride upon the hips; the Klings, erect, handsome, and picturesquely clad in bright and graceful garments; Malays, in many-coloured patterns, chewing betel-nut, and walking as though bound on a pleasure excursion; Hadjis, who have made the pilgrimage to Mecca, with yellow turbans on their heads; the crowds of little children nearly or quite naked, of all shades of colour, and grotesquely shaven; the half-castes, with a sprinkling of Europeans, altogether constitute an *ensemble* which the traveller from temperate climes cannot fail to regard with intense interest.

A great deal is said of the number of tigers which are supposed to infest Singapore. A recent writer* estimates the deaths from these animals at 365 per annum, chiefly among Chinese cultivators of the Gambier plantations in the interior of the island. Residents at Singapore, however, always assured me that there was no danger whatever from them, and that they are never seen now. Indeed the officers at Fort Canning expressed their belief that many of the instances of disappearance of Chinese which were attributed to the ravages of tigers, were really cases of murder by other Chinese; and truly it would not be difficult in distant and secluded plantations for one man to secure the disappearance of another to whom he owed a grudge. I do not, however, accuse them of this. That there are tigers in Singapore no one doubts; but while Mr. Cameron supposes that there are as many as 20 couples in the island, others believe that six or eight tigers would be a sufficiently high estimate. Formerly they came much nearer to the town, but still it is long since the jungle near plantations was infested. A gentleman who has lived all his life in Singapore, and had possessed large plantations in various places, assured me that the only occasion on which he had been alarmed by a tiger was many years ago, when he was residing on a plantation about three miles from the town. One night he was startled by a very peculiar, terrified cry from nearly under his window, which he threw open, and found it proceeded from a watchman employed about the premises, who could only exclaim, "Tiger! tiger!" On being questioned, he declared he had seen two eyes like

* Our Malayan Possessions in Tropical India, by John Cameron, F.R.G.S.

glowing coals shining upon him through the darkness. He was laughed at as a dreamer, but persisted in his story, and next morning the tracks of a tiger were very apparent about the spot.

The Gambier plantations are very thick, the plants being placed very close together, and growing to the height of five or six feet, so that they form a jungle in themselves; and there is abundance of rank grass, which affords ample cover for tigers; but were so many persons killed as asserted, no Chinese could be induced to go and work in them. A very old resident in Singapore told me that he had long been in the habit of rambling about the jungle for weeks together, often penetrating five or six miles from the government roads, and yet he never saw or heard a tiger, though he has seen their tracks in his plantation. It is difficult to conceive what can induce tigers to cross over to Singapore; for although there are a few deer and plenty of pigs in the island, there is a much greater variety of game in the Johore peninsula. Can it be a taste for human flesh, which is more plentiful in the island?

An old guide-book stated that so numerous were the tigers that on the arrival of the steamers the passengers might see them come down to the water's edge to drink (salt water, of course). And it was a common statement that the island was *infested* with tigers. But, at the present day at least, one might walk all through the island without seeing a trace of them; and the *roads*, at all events, are perfectly safe, not only from tigers, but also from robbery or violence.

At Singapore, and also at Labuan, the little luminous beetle, commonly known as the fire-fly (Lampyris *sp. ign.*) is common. When flying singly it shines with an intermittent light, which alternates with darkness; but on fine evenings and in favourable (that is, damp and swampy) localities, they present a very remarkable appearance. Clustered in the foliage of the trees, instead of keeping up an irregular twinkle, every individual shines simultaneously at regular intervals, as though by a common impulse; so that their light pulsates, as it were, and the tree is for one moment illuminated by a hundred brilliant points, and the next is in almost total darkness. The intervals have about the duration of a second, and during the intermission only one or two remain luminous. To all appearance they are not on the wing at the time, but settled upon the tree; for I was able to recognise certain points of light which I especially noticed, and which remained in the same situation at each successive flash. When I disturbed them under such circumstances they flew about at random, each one giving out a more rapidly intermittent light. At Labuan, however, I have frequently seen them shine with a steady light as they flew along, looking like little falling stars of the second, or even first, magnitude.

The candle-fly (Candelaria), with a curved and pointed head, does not appear to give out light as its name would indicate. These insects are found in Labuan and Sarawak, and frequent the upper parts of the lofty trees of the jungle. The only *fire-fly* of these parts is the above-mentioned little beetle.

The Agri-Horticultural Society's gardens are situated about three miles from town, along Orchard Road, one of the prettiest outlets of Singapore, with a shady grove of trees on either side for the greater part of the way. They are commonly known as the Botanic Gardens, and when I visited them I expected to have found them something of that character. But they are merely pleasure grounds, kept up by a subscription among the merchants, &c., which does not amount to more than one hundred dollars a month; and they are superintended by a Scotch gardener, who went out some 15 years since, in the time of nutmeg planting. The garden is pretty, but very exposed and unsheltered, and far too hot to walk in, except in early morning or in the evening, when people drive out from Singapore. It is situated on the edge of the jungle, and some strips of the original wood remaining uncut form a pleasant shady walk, while here and there some of the jungle trees have been left standing. Besides these, some fine Coniferæ, as the Norfolk Island Pine, Araucarias, Casuarinas, &c., flourish, and afford a diversity. Many roses are cultivated, and other horticulturists' flowers, but nothing interesting in a botanical point of view, except perhaps the great spider-orchis (Grammatophyllum), a magnificent bush, crowded with its mottled rich brown flowers. A grass is sown which makes a good turf, and a large sheet of ornamental water is in preparation, which will form a very desirable and pleasing feature. The gardeners employed are nearly all Malays.

One of the commonest roadside plants of Singapore is the sensitive plant (Mimosa sensitiva), which grows in profusion in waste places, and on banks by the wayside. It is a very low, spreading plant, of suffruticose habit, seldom rising higher than the grass among which it grows, or more than six inches from the ground, but covering large spots, which are distinguished from the rough herbage by its neat, regular foliage. It seems to be almost constantly in flower, for in October, November, and May I noticed numbers of the little round tufts characteristic of this acacia (Mimosa), and of a pale flesh-colour. The manner in which the aspect of such a little bush is altered by a touch is very remarkable.

Brush your foot over the luxuriant little plant as you pass by, and the whole bush seems to disappear, and you look back for it almost in vain; the leaves have all closed up, and the stems become depressed, and nothing is left but a few withered sticks upon the grass. Try to pluck a spray, and it fades between your fingers; so that it is very difficult to gather and examine it in an expanded condition. But if you will carefully take between the finger and thumb the pulvinus, or swelled base of the leaf stalk where the little thorns are situated, without touching any other part, and pinch it hard before attempting to break off the spray, the pinnæ will remain expanded; relax your firm hold, however, and they will immediately begin to close up.

The Klings here have a mode of obtaining small birds which might prove useful to the practical ornithologist. I have more than once seen one of them beneath a Banyan, armed with a straight tube, or sumpitan, about six feet long, and a piece of soft clay, from which, having broken off a morsel, he rolled it into a little ball between his hands; then, placing it in the tube, and taking aim at a small bird singing in the branches above, he noiselessly blew the pellet, and down fell the bird to the ground. At first, I presume, it was only stunned, or it might be killed; but a companion always picked the bird up and proceeded immediately to cut its throat and place it in his pouch,—not because it was necessary, but because it is against their religious principles to eat animals which have died a bloodless death. A little practice one would imagine would enable a performer to play upon this instrument, not wantonly we would hope, but for the purpose of procuring small birds when they are required for preservation without injuring the plumage.

Singapore enjoys the reputation of being a very healthy

place, and although close to the equator, it possesses a temperature much more moderate than that of many places in higher and lower latitudes. The thermometer seldom varies more than 20° or 22° during the whole year, never rising much above 90°, or falling much below 70°. The nights especially are very cool and refreshing, and enable people to sleep without difficulty, which is one great secret of its salubrity. During the time I was at Singapore in October and November, a sharp squall passed over every night, usually about 2 A.M., accompanied with heavy rain. One night when the squall had been heavier than usual, and it felt quite chilly towards sunrise, I found the thermometer standing at 75°. The following day at noon it was hot, close, and somewhat oppressive; but the thermometer had only risen to 82°, and in the evening at 9 P.M., while it stood at 80°, the air was warm, not close, but comfortable. Knowing that a great star-shower was expected on the occasion of the appearance of the November meteors, I anxiously looked for the day; but the 14th November, as well as a day or two preceding and following, turned out cloudy, so that absolutely nothing could be seen.

Among the numerous islands about Singapore, there is doubtless a wonderfully rich and unexplored region for the marine zoologist, who would find endless occupation for the dredge, as well as by ransacking the coral patches which occur in some places, and are very shallow at low-water. Although my opportunities were very limited while at Singapore, the magnificent species of Comatula (or feather-star) which came under my notice, as well as several new Crustacea of the genera Alpheus and Galathea, proved how well a systematic search would repay the observer with novel and interesting species. The best and most promising shore lies west of the town, and is covered with loose stones, upon which grows abundantly a species of Keramidia, an Alga allied to Jania; while upon these stones I met with seven species of Planaria, of very various and beautiful forms, and all probably new, as well as one species I had already found at Labuan. Nudibranchiates were rare, and several explorations only yielded a single species, a very beautiful rose-coloured one, probably Doridopsis rubra, which occurs among the Indian nudibranchs of Sir Walter Elliot, as well as among those described from Ceylon. Several species of Ascidian Tunicates, small hirsute Crustacea, and Peronias, but no Echinoderms (star-fishes or sea-urchins) of any kind. Among the zoophytes I detected a specimen of the beautiful dendriform Actinia, already described from Pulo Enoe, so that Singapore must be added as a locality for the interesting sea-anemone.

The harbour of Singapore was sometimes beautifully luminous at night, as described in another chapter, and on such occasions abounded with Noctilucæ, which appeared to be the cause of the phenomenon. Sharks from time to time venture among the shipping, probably enticed by the garbage thrown overboard from the vessels, which also attracts a number of large hawks, known under the common name of Bromlykites, which are constantly hovering about, and darting down, seizing some floating mass in their claws. The men of the "Pearl" man-of-war, which lay further out than we did, two or three times hooked a shark during our stay, the carcase of which was gladly purchased by the bumboatmen for a couple of dollars; but we, who were nearer in-shore, never succeeded in taking any of these monsters, although the tell-tale fin was more than once seen not far to seaward.

CHAPTER XVI.

CULTIVATION IN SINGAPORE.

Climate of Singapore—Soil—Nutmeg Planting—Appearance of the Tree—Over-Manuring—The Nutmeg Disease—Its Causes—Ruin of the Planters—Occasional Spontaneous Recovery—Cotton—Coffee—Cinnamon—Sugar-Canes—Gutta-percha—Gamboge—Gambier and Pepper—Fruit Trees—Cocoa-nut—The Cocoa-nut Beetles—Sago Plantations.

The cultivation of the soil in Singapore Island has been carried on with great industry and enterprise, and for a while with success; but unfortunately, after hundreds of thousands of dollars have been spent upon it, the planters have learned, too late, that neither the soil nor the climate of Singapore are favourable to the growth of those productions, such as nutmegs, cloves, cotton, sugar, coffee, &c., upon which such vast sums have been expended and ultimately swallowed up, bringing their proprietors in many instances to ruin.

The climate of Singapore is very peculiar, and is marked by an absence of seasonal change, which, however beneficial it may be to man, has an evil influence upon plants. There is no regular recurrence of summer and winter—no distinctly dry season and wet season, but a remarkable equality all the year round; added to which, the rains, instead of coming at definite periods, are capricious in their fall, and therefore defeat the prognostications of the planters. The temperature does not vary more than 20° or 22° during the whole year, ranging between 70° and 92° as a rule, and not therefore in excess during the hottest seasons. Rain falls upon half the days of the year, neither so frequently nor so heavily now as it did before the jungle was cleared away from the neighbourhood of the town; but the total amount of rain is moderate.

The soil is poor, and will grow nothing without care and plenty of manure. It consists of a fine, compact, reddish clay, in the interior of the island (not having much substance), and mixed with sand, which increases in quantity near the sea-beach, the clay predominating inland, and the sand near the coast. The island was, of course, originally covered with jungle, but there has been a great mania for clearing, and it has been done in an indiscriminate manner, so that no judicious spots of shelter have been left standing, which would have proved invaluable as protection for certain crops, as well as being useful in other ways. The virgin soil, covered with a thin layer of decaying vegetable matter was rich enough; but when, after a little time, its material was exhausted, nothing but plenty of manure would induce the growth of remunerative crops.

Foremost among these crops was the *Nutmeg* (Myristica moschata), a plant which once promised a harvest of prosperity to the settlement; but which, after for a few years producing every result that could be desired, was destined to end in utter disappointment, and, in too many cases, in utter ruin to the proprietors. The nutmeg-plantations of Penang preceded those of Singapore, and were for some years in the hands of the East India Company, who, after expending considerable sums upon them for some years without receiving an adequate return, finally gave them up in disgust, and ordered them to be sold. Taken up by enterprising planters, the Penang spice-plantations for a time yielded ample returns, owing rather to the care which had been spent upon them by the previous possessors. Singapore became a British settlement in 1824; and in the infancy of this settlement it was not attempted to vie with Penang in cultivating these expensive plantations; but about 1837 an impetus was given to nutmeg-cultivation in Singapore with results so promising that everything gave way to the mania for planting this species. Large clearances in the jungle were purchased from Government at considerable distances from town, and expensive bungalows were erected upon such estates, and surrounded by plantations of this valued tree; and nearer the settlement, private gardens were turned into nutmeg-nurseries, and the houses were closely surrounded with nutmeg-groves.

The nutmeg-tree is, when in health, a handsome bushy tree, between 20 and 30 feet high, with numerous dark-green shining leaves. It is evergreen, and ever-flowering, so that fruit and flowers constantly coexist upon the tree— the flowers small, yellowish, and urceolate, and the fruit needing no description here. Being diclinous, a great inconvenience arises from the fact that a great many male trees are planted and cultivated, being undistinguishable from the female trees until the flowers appear. Such trees are of course useless, since they do not bear—*one* male tree to about *twenty* females being sufficient for the purposes of impregnation, and to ensure the swelling of the ovule.

The trees were not allowed to be left to the natural powers of the climate and soil, but were richly manured and forced into yielding heavy crops. To the manner of doing this, and to the extent to which they were forced into luxuriance, may probably be traced the catastrophe which eventually blotted out nutmeg-cultivation from the settlement. Around each tree, and just level with the outer branches, a trench was dug about one foot deep and one foot wide, and this was filled with a manure of cow-dung. The result of this universal treatment was that the trees for a time grew luxuriantly, and yielded large returns. About six hundred nuts, or 8 lbs. weight, were yielded by a good tree during the year; and as the crop was yielded all the year round, independently of season, some plantations produced a picul (133 lbs.) per diem on an average—the value of the picul being 70 or 80 dollars—or from 25,000 to 30,000 dollars per annum.

For upwards of twenty years the planting was carried on vigorously. Plantations changed hands at very extravagant prices; and much money was made during that period. In the year 1860, however, a sudden destruction came upon the trees from an unknown quarter; and, to the dismay of the planters, the trees, which up to that time had yielded magnificently, were attacked with a blight, whose destructive effects could not be arrested, while the source of it defied all inquiry. In the night a tree would be attacked, and the morning light would show its topmost branches withered; the leaves fell off; the disease slowly spread downwards, chiefly on one side of the tree; and, in spite of every attempt to check it (the lower portion often being for a long time green and bushy), the tree became an unsightly mass of bare and whitened twigs. Most trees were entirely stripped in time, and became mere skeletons. Large outlay was expended in the endeavours to arrest the destruction, but it was all thrown away. No situation was exempt from its ravages—hills and valleys alike suffered, nor could any principle be traceable in its promiscuous attacks. Upon a close examination of the diseased parts, it is found that the formative layer inside the bark dries up and turns black; the leaves then wither and fall off; and soon the bark is found to be full of small perforations; but no insect of any kind has ever been discovered in connection with the change, nor has any fungus been charged with the destruction. Its nature has been a mystery and a puzzle with the planters, who have, for the most part in vain, sought for a cause, either near or remote, and whose efforts to arrest it have proved entirely unavailing. I have heard various suggestions offered, some of them of the wildest character, to account for the disease. That which Mr. Josè d'Almeida proposes is by far the most reasonable, and in fact commends itself to the judgment of the vegetable physiologist. It is that the trees had long been unnaturally forced, by digging trenches too closely around their spongioles, and by too rich and long-continued manuring, by which heavy crops, it is true, were for a time obtained, but which at last exhausted the tree, so that the premature decay, thus brought on by inflexible physiological laws, was incapable of being arrested by any after-treatment.

In conversation with a gentleman who once cultivated nutmegs on a large scale, I was assured by him that he could distinguish at least two forms of disease. In one of these it was deep-seated and radical. In many trees which he cut down for the purpose, he found that the central part of the main stem was turning black; and this gave the first indications of the onset of the disease, which was soon followed by the falling off of the leaves and the whitening of the branches.

With regard to the other form of disease, he distinctly

traced it to the attacks of what, from his description, must have been a small black aphis, which perforated the branches, and caused them to wither one by one. I find no two accounts to be precisely alike in respect to the manner of falling away of the trees; but all agree that their destruction was rapid, certain, and irremediable.

When it was found that, in spite of care and lavish expenditure, the trees surely died, a reaction took place. The planters abandoned the plantations in disgust, in many cases while there were still numerous healthy trees; and the land reverted to the Government. In other cases, where expensive bungalows were built upon the estate, they were sold for a small proportion of the sums expended in building them, since they were, as a rule, too far from town to command any competition, and ceased to be conveniently situated. Many planters, both English and Chinese, whose whole estates were invested in nutmeg-plantations, were thus reduced to ruin, and became absolutely penniless; and distress and disappointment everywhere prevailed.

It is a curious fact that many of these abandoned trees, around which has now sprung up a thick jungle undergrowth, have, since they have been thus neglected and left to themselves, *recovered*, and relieve the generally dismal prospect of bare branches and skeleton trees. I have myself seen these dark-green healthy trees in many situations where they are quite uncared for, even amongst the oldest plantations in the island; and this fact seems decidedly corroborative of the idea that the disease was one of exhaustion and decay, arising from unnatural forcing. Another fact is significant, viz. that, while at Penang, where this cultivation, as described, was carried on with the greatest vigour and the greatest expenditure, the destruction has been most complete and marked, at Malacca, where the people were not so rich, and could not afford to manure the trees so highly, they have not suffered so severely as at Penang and Singapore.

At the present moment there is no such thing as nutmeg-cultivation, either at Penang or Singapore; nor does it seem probable that the experiment will be again tried. Planters are now persuaded that neither the soil nor climate is favourable for their production; and, as we shall presently see, other crops have fared but little better. The trees which still exist are neglected and abandoned by their owners, though they still yield nutmegs. These are gathered by any Chinese or Malays who take the trouble to do so; and the few nutmegs, insignificant in quantity, which now find their way into the Singapore market, are obtained in this way,—a clear gain to those who carry them there.

Cotton (Gossypium herbaceum) is another product the cultivation of which has been attempted in Singapore. The cotton-plant always thrives well in private gardens; and I have seen large pods of good quality on plants in such situations. The only large plantation which has given it a fair trial, however, was that of the late enterprising Mr. d'Almeida, who for two successive years expended considerable sums on the experiment. But cotton cultivation failed for the same causes as those above referred to—the absence of regular seasonal changes, and the irregularity of the downfalls of rain, which cannot be predicted with any certainty, and therefore cannot be guarded against. The cotton grew magnificently; the pods were produced and burst open, and then a down-pour of rain would ruin the fibre before it could be gathered. Another cause which led to its abandonment was the appearance of a small red beetle, which proved very destructive to the pod.

The same gentleman made a trial of planting Coffee (Coffea Arabica), and spent and lost many thousands of dollars by the unthankful experiment. It has also been attempted by others without success; and a company formed for that purpose failed. Here again the causes of failure are chiefly natural ones, of the same kind as those already alluded to. The coffee-plants require shelter; and the indiscriminate cutting down of the jungle had left the country entirely open, and no shade could be obtained. Then the irregularity of the seasons prevented the plants from attaining that perfection which otherwise they might have done, while the uncertain rains were a further source of injury to the crops. The flowers might be in promising profusion, when a heavy shower would suddenly fall upon them and destroy two-thirds at one blow. Another difficulty which interferes with this and other cultivation is the comparatively high price of labour. Anything which requires much manual labour in the preparation is sure to languish at Singapore from the difficulty of persuading the Malays to work for any consideration; and the Chinese are the only people who can be induced to undertake laborious occupations.

This last cause has been mainly influential in preventing the cultivation of Cinnamon (Cinnamomum zeylanicum). This tree, with very little care, grows beautifully in Singapore, and would doubtless prove a source of wealth were it not for the great expense of the manufacture. The various and tedious processes which the bark has to undergo in its removal and preparation cost more than the spice will fetch in the market. In other cinnamon-producing countries, as in Ceylon, these processes are performed chiefly by children, who, of course, are paid at so low a rate as to render the preparation remunerative; but in Singapore the population is not large enough for this; and expensive adult labour only is procurable, and that with some difficulty.

The Sugar-cane (Saccharum officinarum), on the other hand, has failed from natural rather than economic causes. The chief obstacle to its cultivation is the poorness of the soil, which can only be remedied by adding plenty of manure; and when this source of additional expense is added to the high price of labour, considerable margin is subtracted from the profits. Still, with abundance of manure, the sugar-cane thrives extremely well; but now another natural cause steps in and neutralises the result: this is the rain, the uncertainty of which, or rather the constancy of which, is a serious obstacle. The saccharometer, instead of registering 11° in the sweet juice, is sometimes reduced to $7\frac{1}{4}°$ after rains, which appear to dilute the sap and deteriorate the produce. In a plantation ready for cutting, perhaps 50 acres may be got down one day and of good quality; and then a heavy rain comes before the rest can be cut; and this proves to be of considerably inferior quality.

The late Mr. d'Almeida was the first to call the attention of the public to the substance now so well known as Gutta-percha. At that time the Isonandra gutta was an abundant tree in the forests of Singapore, and was first known to the Malays, who made use of the juice which they obtained by cutting down the trees, and which, when collected, they boiled and purified. Mr. d'Almeida, unacquainted with England and its institutions, and acting under the advice of a friend, forwarded some of this substance to the Society of Arts. There it met with no immediate attention, and was put away uncared for. A year or two afterwards Dr. Montgomery sent specimens to England, and, bringing it under the notice of competent persons, its value was at once acknowledged, and it rapidly became an important commodity. In any case it was introduced from Singapore; and the sudden and great demand for it soon resulted in the disappearance of all the gutta-percha trees in Singapore island. The forests of Johore, however, yield a vast supply; though these must fail in time, when it is borne in mind that to abstract the juice the tree is always cut down, the produce of a single tree averaging 11 or 12 lbs.

With regard to Gamboge (Cambogia gutta), it has never been regularly cultivated in Singapore. The late Mr. d'Almeida, already referred to, introduced some trees from Siam, but simply as a matter of curiosity and for experimental purposes. These trees have not been protected in any way, but nevertheless they thrive well; and the soil evidently is well suited to them. The plantation in which they were placed has changed hands, and no care has been taken of the trees; but those I saw were green and flourishing, bearing abundance of flower and fruit, and yielding, upon the slightest incision, an abundance of yellow resinous juice. In their immediate neighbourhood are numerous healthy seedlings springing up uncared for; and I was assured that the seeds carried by birds have been taken to spots at a distance from the trees originally planted; and one of the largest and healthiest trees I saw was pointed out to me as one which had grown there spontaneously, and probably owed its origin to this cause. I preserved specimens of this tree, and of the female flowers in spirits, as possessing especial interest for the pharmaceutical botanist. But, although to all appearance it would do well, no one has taken up the matter of cultivating them, and the existing trees are quite neglected. For this reason also I was unable to procure any specimen of the gamboge produced by them, though I was informed by the Chinese gardener who showed me the trees that incisions were made in this bark, and small bamboos were applied to the incised spots to receive the juice. Hence the *Pipe-gamboge* of commerce. I may add that the soil on which the Gamboge appears to thrive so well is a reddish sandy soil, containing a little clay, but a larger proportion of sand.

This brief account of the past cultivation of Singapore would not be complete without some mention of two plants which have been largely planted by the natives, though the cultivation of them is now on the decline. These are Gambier (Uncaria Gambir) and Pepper. With regard to the first of these—gambier—the mode of its preparation demands a very considerable supply of firewood; and therefore it has always been planted in clearances made in the jungles of the interior of the island, and distant from the town. Here the planters squatted, an   for a long while successfully cultivated this favourite .sticatory. The gambier plant is a creeping annual, and rises to the height of six or seven feet. In eight months the young plants are fit to be cut; and the young leaves and shoots are cropped and boiled; and the extract thus obtained is evaporated to a paste, dried, and cut in small blocks an inch square, which are then ready for the market. The workers in these plantations are exclusively Chinese; and the proprietors are also of that nation. The gambier is a plant which very rapidly exhausts the soil; and the quantity of wood required for boiling the shoots demands the immediate neighbourhood of an inexhaustible supply. In course of time, therefore, the wood has all been cut down close to the plantation; and the necessity of having to convey it a mile or so is fatal to the successful cultivation of the drug; consequently, gambier-planting is now fast disappearing in Singapore.

It had always been found profitable to combine with gambier-planting the cultivation of pepper; partly because this could be attended to in the intervals of gambier-cropping, but chiefly because the boiled shoots and leaves of the gambier, after the astringent was extracted, formed an excellent ready-made manure for the pepper, free of expense, which no other manure would have paid. As therefore the planting of gambier declines, that of pepper must necessarily decline also, and as the two rose together so they must also fall together. Considerable quantities of pepper are still produced in Singapore, but not nearly so much as formerly; and many of the gambier and pepper clearances have reverted to the Government. In the peninsula of Johore, however, there are abundance of pepper and gambier plantations.

It may be asked, however, if Singapore has failed in realising the expectations of planters in so many instances, and so many different crops have one by one proved ruinous to their proprietors, what *will* grow remuneratively in the island?—or will anything do so? The answer to this has been solved of late years. In the first place it is found that all fruit-trees flourish in the soil of Singapore; and breadfruit, jack, dookoo, mangosteen, pineapple, plantain, rambootan, custard-apple, mango, guava, and durian, with many others, now occupy the plantations in which nutmegs were formerly grown. The last-named fruit, so great a favourite with some, and so detested by others, is produced in such quantities that 50 dollars are given for the produce of a single tree.

But the one tree in which is now centred the promise and the hope of the Singapore planters is the Cocoa-nut (Cocos nucifera). It does not appear to be indigenous, for none are found in the jungle; but it was long since introduced by the Malays. It is comparatively of late years, however, that European planters have looked upon it as a source of wealth, and foreseen that it may prove in course of time to be the most important production of Singapore. The original cocoa-nut plantations are yielding golden returns; and within the last ten years, or less, a great impetus has been given to the propagation of a tree to which the sandy and poor soil of Singapore seems admirably adapted. The trees thrive, and the only drawback is that several years must elapse before they attain such a growth as to yield any recompense for the original expenditure. The uses of the tree are numerous; but it is to the oil that the planter looks for his reward. With proper machinery for separating this oil, the rapidly-extending cocoa-nut plantations bid fair to place cocoa-nut oil in an important position among the exports from Singapore. The cocoa-nuts, however, are not free from their enemies, in the shape of two beetles—one, a large Curculio (Rhynchophorus Sach), nearly as big as the English stag-beetle, and the other an Oryctes (O. Rhinoceros), so called from its projecting horn. The first of them is called in Singapore the *red beetle*, from a blood-red mark upon the upper part of the thorax, and it probably attacks the nut; while the second feeds upon the terminal bud of the palm-stem. When thus attacked, the bud dies, and the crown of leaves falls off, leaving the graceful cocoa-nut tree a mere tall bare pole. Such bare poles I have seen representing all that remains of the betel-nut palm (Areca catechu), which is subject to the attacks of a similar beetle. In Penang, thousands of cocoa-nuts are destroyed by the ravages of these insects. In the cocoa-nut plantations men ascend the trees and examine narrowly for these insidious enemies, which they find in large numbers. They forthwith pierce them with a sharp stick, and passing a string through them, hang them up in festoons at the entrance of the plantation. Such strings of beetles, some dead and decaying, some still alive and kicking their legs about, I have seen in the plantations of the island. At the present moment, however, the cultivation of cocoa-nuts is merely in its infancy; and the exports are confined to places in the immediate neighbourhood of Singapore.

An enterprising gentleman is cultivating the sago-palm on a large scale, about eight miles from Singapore. The plantation (containing at present 10,000 trees) is still young, and will not begin to yield for about five years; but this flourishing state of the trees, with the aid of a certain amount of manure, gives full promise of a successful result. When the trees are ready to cut, he intends to apply machinery to the preparation of the sago; for, according to the present primitive modes of the natives, a man (Chinese) and his wife, their adult son and wife and two children, are employed a fortnight in preparing the product of a single tree.

Let me add, too, with regard to labour, which I have spoken of as comparatively dear—a Malay or a Chinese commands a price of $3\frac{1}{2}$ to 4 dollars a month; while in Java 3 rupees is considered good wages; and, besides being doubly expensive in Singapore, the workman always takes two hours in the middle of the day for rest, and stops work the moment the clock strikes six; while the men are so chary of their labour that it is necessary to have overseers to keep them at it.

CHAPTER XVII.

JOHORE AND THE STRAITS.

Excursion to Tanjong Putri—Chinese Carnival—The Tumonggong—Sing-songs—Chinese Thespians—Gambling Parties—The Game of "Poh"—Gambling in Singapore and Hong Kong—Mountebank Dentistry—Opium Smoking—Statistics of Consumption—Value of Imports—Chinese Opium—Considerations—Saw Mills—Horsburgh Lighthouse—Coast of Johore—Habits of the Pill-Crab—Ubiquity of Ants.

A FAVOURITE excursion from Singapore is that to the back of the island, where is a commodious bungalow, situated on the border of the Straits, which are here not more than half a mile wide; and opposite which is the town of Tanjong Putri, at the southernmost extremity of Johore. When I visited this place, it was in an unwonted state of excitement, from the fact that his Highness the Maharajah of Johore was visiting his residence there for the first time since his return from England, where he had been received with very great distinction, having among other honours been installed a Knight of the Star of India. The town was like a fair, and the Chinese especially were busily employed in turning the occasion to advantage. Gambling places and sing-songs were driving a great trade, and the juggler and mountebank were in their glory.

The shores on both sides of these Straits, between the Island of Singapore and the Malacca Peninsula, are densely wooded, with here and there a cocoa-nut plantation, having a hut built upon it, and impenetrable mangrove thickets skirting the beach. The water was so shallow that we ran aground, and had to wait for the tide, with only five feet of water under our bows—a mishap which delayed us so much, that it was dark before we arrived; but the bright lights, fire-works, and noise of tom-toms were sufficiently distinct to serve as landmarks to guide us to an anchorage.

The morning light showed that Tanjong Putri was simply a clearance in the jungle at the south point of Johore, with apparently no outlet on the landward side. The noises which we had heard on shore on our arrival still continued, having gone on without intermission all night long—and indeed they did not cease as long as we were within hearing, for the Chinese were keeping carnival.

The occasion was an excellent one for observing Chinese characteristics—for the larger part of the population appeared to be formed of Celestials, although of course the real natives are Malays, who appear to be attached to their native ruler, and to be moreover proud of the travels from which he had just returned, and of the attention which he had received from high quarters in England. I had an interview with his Highness, who is styled the Tumonggong of Johore, and had the honour of smoking a cigar and drinking a glass of sherbet with him. He is a good-looking young man of 30 or 31 years of age, rather stout, and taller than the average of the Malays. Unlike his subjects and countrymen, he cultivates a moustache, and, as might be expected under the circumstances, he wore a European costume. His manners were gentlemanly and agreeable, and he treated me with unaffected urbanity and good-will. He speaks excellent English; and the conversation naturally turned upon his late visit to this country, which he seemed to have greatly enjoyed; and doubtless the new and enlarged views which he has imbibed from such a visit, and the attentions which he received while here, cannot fail to have a beneficial influence upon his Eastern rule. The Maharajah has become invested with an importance and interest in the eyes of the Malays which he could hardly have otherwise acquired at home; and they seemed to vie with one another in showing their loyalty and service. He is, moreover, indebted for his present position to the policy of the English Government, who transferred the rule from the former Sultan to his admiral, the father of the present prince—an act of Sir Henry Butterworth which has been freely canvassed, but was doubtless justified by State reasons which that Governor could well appreciate.

The sounds which had greeted our arrival at Tanjong Putri, I soon discovered arose from a Thespian entertainment, under the auspices of the Chinese; and inasmuch as all this class of performances had a very great family likeness wherever I had an opportunity of witnessing them, I may say a few words descriptive of the singular character of this exhibition.

There were two of these *sing-songs*, or open-air Chinese theatres, which were centres of general attraction, placed, however, almost side by side, so that the proceedings of one thrust themselves upon the spectators of the other, and somewhat marred the effect of both. They were good types of Chinese theatricals, and consisted of spacious stages, open in front, and erected above the level of the heads of the spectators, with *attap* coverings for the benefit of the performers, but nothing of the kind for the lookers-on, who either stood sweltering in the sun, or, if they preferred it, took shelter under the verandahs of the shops on the other side of the road. At the back of the stage, in the centre, was placed a table, behind which were the musicians, some hammering upon tom-toms of various sizes, which gave out a more or less resonant sound, others playing upon the fifes, and producing sounds which might readily be mistaken for bag-pipes. Besides this there were three embroidered mats hanging down behind the stage, and these together constituted the scenery, properties, orchestra, and all equipments which their Thespian simplicity required. At the back of the stage a door on either side served as an entrance and exit for the actors, who always came in at the left hand and retired at the right. The play appeared to be a burlesque, and the actors used the burlesque movements of the low comedians on our own stage, only more coarse, clownish, and exaggerated. They were men and women in this case, though more commonly the women's parts are performed by men in female costume. The men were dressed in the highly embroidered robes and painted grotesque masks which are familiar to every one who has turned over rice-paper picture books; and the women spoke in a high falsetto voice, quite different from the female treble. They came in by the left door in small parties, flourished about, and shouted, passing slowly in front of the stage, and then disappeared on the right side, and were succeeded by another party, the same party again re-appearing after a short interval. There seemed to be no termination to the story, nor any limits to the endurance of the actors or spectators; for the latter kept up a constant crowd in front of the stage, behaving, however, with great decorum and even gravity, and showing little inclination to laugh at the antics of the players; and I could only judge of the actors' endurance, from the fact that the accompanying noise of tom-toms and fifes ceased not day or night all the time we were within hearing.

There was the usual mixture of barbarism and splendour which characterises all Chinese ceremonials. The sides of the stage were occupied by a number of dirty, half-naked boys and men,—regular *gamins*,—who perched themselves upon the stage itself by some peculiar right, by virtue of which they seemed entitled to *reserved* seats; and the actors themselves exhibited strange contrasts to their richly embroidered and really handsome robes, for these were usually open in front, disclosing their brown, bare skin from the neck to below the navel. I found it impossible to gather any hint as to the nature of the story or plot of the play.

Among the amusements of the Chinese population at such a time of festivity, gambling holds a very prominent place. The Chinese are passionately addicted to this vice, and spend days and nights over cards and dice, imbibing the passion from their very earliest years. A child who has become the possessor of two cash, and goes to invest it in sweetmeats, will either gamble it away before he arrives at the stall, or will toss the vendor double or quits while he still holds that vast sum in his hands. On the present occasion there was in the town a large covered area entirely occupied by gambling parties. Each party occupied a small square space, upon which a piece of carpet is spread, and around which the players squat upon the ground, three or four being engaged in counting out the small change for stakes, and attending to the business of the bank, while space was afforded for about six more, always Chinamen, in dress and appearance indistinguishable from coolies, who kept up a constant rattle of money and dice. All the available space between the mats was occupied by standing spectators, who not unfrequently joined in or filled up the places of those who left.

There are several methods of gambling employed, and it is not easy for a mere bystander to catch the spirit of the game; but most of them are very simple. Thus, for example, a board is produced with twelve squares, and the stake is made on one of them; if that square turns up, the lucky depositor receives twelve times his stake. Or there is another board, upon which are painted representations of 36 different animals: on one or more of these a stake is made, a successful hit winning 36 times the stake. A third method is as follows:—the keeper of the bank takes up a handful of coin, and a board is produced, divided into four squares, marked respectively 1, 2, 3, 4, on either of which a stake is made. The coins are then counted until four, or less than four, are left. If there are just four, the man who has staked on No. 4 square wins; if two, No. 2 succeeds and gets the handful; otherwise, the stake is forfeited.

But perhaps the most favourite game with the Chinese is that called "Poh." This game is played with a single die, and a small, solid brass box, in the upper part of which is a square hole in which the die fits. Each face of the die is half red and half white, and is inscribed with Chinese letters. The die having been shaken by the banker in a red bag, he takes it out, and, without looking at it, places it at once in the box, and covers it with a brass lid. Giving the box a spin, the players stake their money upon the colour, placing it on that side of the box on which they expect that colour to be. If one betting on white places the money on the red side, it is of course lost, but if on the white side, he wins double stakes; if on the side on which the colours are divided, he loses. They sometimes stake on the corner of the die, in which case, if three white or three red halves of the die on the three visible sides meet at that corner, white or red wins a single stake; but if the colours are mixed, neither can win.

Besides these small gambling-places, in which, for the most part, the lower classes amuse themselves, and in which they were clustered like bees around some fifty banks, there was a regular "hell" near by, at which the more wealthy classes, chiefly from Singapore, played to their hearts' content far into the night at the same games, but for higher stakes. Here one man lost on this day 7000 dollars; and some Chinamen who had been "cleaned out," expressed their intention of sending to Singapore the following day for large sums, with the avowed intention of breaking the bank. By large sums they meant, say 5000 dollars, and they would play until they lost it all, or fulfilled their threat.

Up to the year 1829 gambling was permitted by the government of Singapore. The gambling-houses were farmed, and from 1820 to 1829 the revenue from this source had increased from 5,725 dollars to 33,864 dollars. It was then abolished, and fines are now collected in the magistrate's court for breaking the law in this respect. It is said that the fines collected during the first four months of 1864 amounted to 6,112 dollars, or £1,370. I was informed that since the prohibition, gambling has been exceptionally permitted for some days at the time of the China new-year, when not only the Chinese population, but many of the leading merchants may have been seen eagerly mixing with the speculating crowd, and winning or losing with the rest.*

One other element of the busy and motley scene may be mentioned—viz., the mountebank dentist. He was a Chinese, and standing in a public place, loudly invited patients to be relieved of their troublesome teeth. Several came forward, and the treatment was not a little singular and puzzling. Clapping a red plaister upon the cheeks, over the spot where the guilty tooth was situated, he, at the same time, put inside the mouth a small quantity of a kind of white paste. Then inserting an instrument which looked something like an ordinary dentist's key, he rapidly whipped the tooth out entire. But the most curious part of the circumstance was that no cry escaped the patients; and on narrowly watching their features, not the slightest symptom of momentary pain was revealed. But the bleeding fangs of the teeth as held up to view negatived the idea that there was any trickery or delusion. The price of the operation was only 10 cents (5d.)! Sometimes the fellow pretended to charm the tooth out without any operation—a feat which he accomplished by sticking the plaister on the face, and inserting the white paste within the mouth as before, after which, instead of using any extracting instrument, he stuck against the tooth the pointed end of a piece of folded paper containing a little of a black substance which looked like pitch. Then having kept the patient waiting for three or four minutes with his

---

* Since this has been written gambling has been legalised among the Chinese population of Hong Kong—a step which while it has naturally given great offence to certain European classes, will be regarded leniently by those best acquainted with Chinese character, and will save the police a vast amount of trouble in hunting up and bringing to justice the numberless cases in which the attempt to restrict this Chinese institution was constantly being evaded by all classes, in whom the habit is too much a second nature to be eradicated by legislation.

mouth shut, he would tell him to cough, when out came the tooth with no further difficulty. I know not what jugglery was used, but these effects were presented to the eyes of attentively watching bystanders.

Among the phases of Chinese dissipation incident to scenes such as I am describing, of course opium-smoking has its place. This, however, is a subject upon which the opinions of some in this country, who are unacquainted with real facts, are so strong, and their feelings so excited, that it seems desirable to give some trustworthy information which may guide them to a proper appreciation of the true extent of the evil, and may enable them to compare it properly with that vice in this country to which it bears most resemblance—viz., drunkenness.

It is a common idea that opium-smoking obtrudes itself upon the notice of every traveller in China, and that the debasing and destructive effects of it meet the eye at every turn. This is, however, a great mistake. Opium is an expensive luxury, and the supply, which is equal to, and regulated by, the demand, is very limited. Like all other luxuries it is doubtless liable to abuse, and no one will attempt to deny that, like spirit drinking, it is sometimes carried to excess; but the cases of confirmed opium-smoking in China bear no manner of proportion to those of excessive drinking in England. This can be easily proved by a reference to the statistics of the opium-market of Hong Kong, through which all opium except that of native manufacture must pass. A person not conversant with the value of the drug is surprised to learn that a *chest*, which contains $133\frac{1}{3}$ lbs., or one picul of opium, is worth about £150.

As a general rule a man smokes about 5 mace of opium at a time, or we may say 5 mace per diem for an ordinary opium smoker. This amount, multiplied by 365 (days), makes 182·5 kandareens, or 18·25 taels (ounces) per annum.

In round numbers, therefore, an ordinary smoker consumes 20 ounces per annum. And since a chest of opium contains 133 lbs., it will require 106 persons smoking at this rate to consume one chest in a year. Now the annual consumption of imported opium is 100,000 chests, and it is believed that about the same quantity is also manufactured in China. This will make a total of 200,000 chests demanded annually by opium-smokers of all classes and degrees.

The extreme value of imported opium one year with another is 700 dollars per chest, and Chinese opium is very much cheaper. At the steady rate of consumption indicated above, viz., of 106 persons to one chest of opium, it would cost the consumers $6\tfrac{3}{4}$ dollars (about 30s.) per head per annum. Supposing, however, that this consumption and expense were spread over the whole population of 300 millions, it would amount to less than a quarter of a dollar, or about one shilling per head per annum. But this estimate must be still further lowered by the following considerations—viz., 1st, that Chinese opium is produced at only one-fifth of the price of the Indian drug; 2nd, that of the Indian opium 6,000 chests are annually diverted to the Straits settlements, Borneo and the neighbourhood, while an unknown quantity goes to Australia, California, &c.; and 3rd, that of the raw opium a considerable per centage is lost in the preparation of the drug for consumption. Thus of Malwa opium, 30 per cent.; of Patna, 35; of Benares, 35; of Persian, 20; of Chinese, 20; and of Turkey no less than 42 per cent. is waste, the remainder forming the real extract for the smoker.

Again, although the moderate estimate of five mace per diem is correct for the mass of opium-smokers, there are doubtless some who abuse the indulgence, and both spend and consume much more—thus still further reducing the number of consumers, who must really form a very small proportion of the entire population, viz., about twenty out of the three hundred, millions, or one in fifteen; that is, 6½ per cent. These are principally the sea-board population—the inland people being for the most part unacquainted with the drug.

The Chinese produce a large quantity of opium, the exact amount of which is hardly known; but it is not valued by them as Indian opium is. It is strong and pungent, and bites the tongue, producing a maddening effect when taken in excess—and bearing the same relation to Indian and other imported opium that strong brandy does to mild wine. The apologists for opium importation affirm that the Chinese *will* have the drug, and did they not import it in a mild and comparatively harmless form, the opium-smokers would use all the more of their own inferior and intoxicating substitute. The Chinese Government derives a revenue of 50 dollars per chest on imported opium—or rather, should do so, but this tax is for the most part evaded.

That opium-smoking is a vice, and leads to evil, is not for a moment to be denied, but that it is of that extent which is commonly believed by some philanthropists well-disposed but ill-informed, is evidently a mistake. Undoubtedly if it could be rooted out of the customs of the Chinese people, it would be a desirable end—and so it would be if drunkenness could be eradicated from the English people—but both ideas, we fear, are equally Utopian and Quixotic. Merchants engaged in the opium trade are loudly condemned; but, to be just, those engaged either directly or indirectly in the production of ardent spirits should meet with an equal amount of reprobation. But there can be no manner of doubt that drunkenness is far more productive of misery and crime in this country than is opium-smoking in China, while we are apt to forget the consideration that we are a professedly Christian people, while the opium-smoking Chinese are heathens, with a very imperfect natural appreciation of morality, as understood in the West.*

Before leaving Tanjong Putri, I visited the extensive steam saw-mills, in which a variety of circular and perpendicular saws were at work upon wood of all sizes, from small planks to enormous trunks of trees. These mills are worked by a company, principally Europeans, but in which the Tumonggong possesses an interest. The workmen are all Chinese, who live in a separate village, which is enclosed, and the gate to which is kept locked during working hours. There are similar saw-mills, but on a much smaller scale, at Singapore.

In passing several times up and down the Singapore Straits, the lighthouse on Pedro Branco Island, commonly known as the "Horsburgh Light," is a conspicuous and interesting object. I one day paid it a visit, and rambled over it from top to bottom. It is built on a rock to the east of Singapore, at 28 miles distance, with soundings of 17 to 23 fathoms all the way. It was a lovely morning,

---

* Confirmed opium smokers, it is well known, suffer severely when deprived of the drug, and the vice sometimes assumes a form which is analogous to *dipsomania*. The friends of such persons have occasionally brought them to the European medical men in Canton to be cured, and a cure is not difficult to be effected by proper treatment and supervision; but it of course depends upon the firmness and principle of the patient to refrain from relapsing into the bad habit.

and quite calm, so that I was able to land at some steps cut in the almost perpendicular side of the rock upon which the lighthouse is built. Although the rock, however, is naturally very inaccessible, a sort of movable pier is constructed, by means of which a landing can be effected at almost any ebb tide. The lighthouse, a testimonial to the invaluable services of the author of the "Directory," is a cylindrical building, with a basement and six stories, which are ascended by narrow ladders, to the light-room at the top. This contains nine cata-dioptric lights, arranged in sets of three, movable by clock-work, so that the angle between each set shows dark. The light is visible once in a minute, and is seen 15 miles. The rock upon which the lighthouse is built, is an irregular, much broken, rounded mass of grey and compact granite, extending out northward in a reef, but with only a few rolled stones at the south. It was commenced in 1850 and finished in 1851, and in many respects closely resembles the Bell Rock Lighthouse, 11 miles east of Arbroath. The chief light-keeper is an Englishman, who is assisted by Malays.

On the rocks a number of Grapsi were running about, and a few Ligiæ; but no other marine animals except fishes were visible, although it was nearly low neap tide— the rocks being too smooth and too much exposed to harbour the more delicate species. The leaping-fish (Periophthalmus), of a large size, were pretty numerous, and it was amusing to see them climb up the steep and smooth sides of the rocks by a series of jumps, assisted by a wriggling movement from side to side—so that each time they alighted the tail was strongly curved on either side alternately. Some low black rocks in the neighbourhood looked as if they were covered with snow, but a telescope resolved the appearance into dense crowds of thousands of white birds, whose general movement gave the rocks a quivering aspect, as when the rarefied air ascends from a heated surface.

The shores of Johore, bordering on the Straits, are everywhere thickly wooded, the jungle coming down to the water's edge. The low banks are seldom relieved by a hill, or anything which serves to distinguish one part from another, and not a habitation is anywhere visible. I one day landed upon the beach at South Point, and spent some hours in exploring. The coast was rocky, with reefs of porphyritic stone containing large crystals of albite; and a shelving, sandy shore extended so close to the edge of the jungle, that only a yard or two was left dry at high water. In the jungle, Cycads and screw-pines abounded; and I fancied I could trace the tracks of large animals, which my imagination helped me to believe were tigers, upon the higher parts of the sand. Butterflies of the same type as those I had observed at Labuan were pretty numerous.

A curious little Crab is common upon the sandy beaches everywhere on these coasts. I observed it abundantly at Labuan, and at Singapore and Johore, and other places, where, immediately after the tide has gone down, the smooth beach is covered with loose, powdery sand and holes of various sizes, from such as would admit a small pea to those big enough for a large filbert, but usually of the former dimensions. A closer examination showed that little radiating paths converged among the litter of sand to each hole, and that the sand itself was in minute balls or concretions of a size proportionate to the calibre of the holes. The rapidity with which the shore was covered with myriads of such concretions was very surprising, as at first there appeared no living thing to which they might be attributed. I naturally supposed that the little crab inhabiting the hole had ejected the sand in little balls in the construction of his habitation; but an approaching footstep was an immediate signal for the disappearance of the little creatures. By remaining quite quiet, however, on a patch 30 or 40 feet square, which was covered with their holes, I was able to watch their remarkable habits. On the first approach, a peculiar twinkle on the sand was visible, which required a quick eye to recognise as a simultaneous and rapid retreat of all the little crabs into their holes, not a single one remaining visible. Kneeling down and remaining motionless for a few minutes, I noticed a slight evanescent appearance, like a flash or bursting bubble, which the eye could scarcely follow. This was produced by one or more of the little crabs coming to the surface, and instantly darting down again, alarmed at my proximity. It was only by patiently waiting, like a statue, that I could get them to come out and set to work. They were of various sizes, the most common being that of a largish pea. Coming cautiously to the mouth of the hole, the crab waited to reconnoitre, and if satisfied that no enemy was near, it would venture about its own length distant from the mouth of its hole; then rapidly taking up particles of sand in its claws or chelæ, it deposited them in a groove beneath the thorax. As it did so a little ball of sand was rapidly projected as though from its mouth, which it seized with one claw and deposited on one side, proceeding in this manner until the smooth beach was covered with these little pellets, or pills, corresponding in size to its own dimensions and powers. It was evidently its mode of extracting particles of food from the sand. I made many attempts to catch one before I could succeed, so swift were they in their movements. Preparing my right hand, and advancing it cautiously, I darted it out as rapidly as I could to secure the crab; but it was too quick, and had regained its hole. At length, after repeated attempts, I caught two specimens, which immediately curled themselves up and feigned death. I put one of them on the sand to see what it would do. At first it did not attempt to move; but after a short time, by a twisting and wriggling movement, it rapidly sunk into the sand and disappeared. I had attempted in vain for a long time to cut off one of the crabs from its hole, so that I might fill it up and observe whether it would go into a neighbour's hole, and with what result. But as I could not succeed in doing this (and it was frightfully hot work stooping over the sand under the direct rays of the tropical sun) I put one of the crabs I had caught into a hole already containing a crab; but no result followed. I attempted to dig it up again in vain. I dug up many holes; but though I soon arrived at the soft and wet sand beneath, I never succeeded in procuring a pill-making crab by digging it out. Nor, when I filled up several holes, did any result follow, as long as I had patience to wait.*

These pill-making crabs are gregarious. Many considerable patches of sand were covered with their holes and pellets, some close together, some more sparsely; but other very large tracts in the neighbourhood had not a single hole upon them. They rapidly make their appearance immediately after the tide has

* Mr. Spence Bate writes me as follows: "The Pill-maker is a very curious fellow, and is very remarkable in its structure. I have drawn it, but have not completed my detailed examination of its structure. It is a new genus, which I have named Sphærapœia (from σφαιρα, a pill, and ποιεω, to make)." The same gentleman has further attached my name to this species, which he calls Sphærapœia Collingwoodii.

left the sand, and go on making their pellets until the water returns again. The first ripple washes all their pellets away, and turns their holes into little funnel-shaped pits.

Of all insects none surely are so numerous or so ubiquitous as the Ants, of which there are numerous species in tropical regions, from the small red ants only just visible, to the large black ones (Formica gigas) fully three-quarters of an inch, and even an inch, long, which frequent woods, and which I saw at the back of Singapore island. When one gets fatigued with walking (and the naturalist *must* walk) it is impossible to sit down any where; for if we sit in the sun we get rapidly baked, and if we sit in the shade we either sit at once in the midst of a community of ants of some species or other; or even if we first carefully examine the place, and think we have discovered a spot which is clear of them, we shall inevitably find the busy insects walking over us in a few minutes, probably brown ants half an inch long, armed with formidable pincers, which they will freely use without waiting for provocation. Even upon the sandy beach, where we might suppose ourselves free from such persecution, the ants follow, bent on foraging expeditions. Thus, on the shore at Johore, I observed large biting ants of a light brown colour swarming about below high water mark; and on the upper parts of the sands, among the drift, I have frequently remarked them.

It is a common circumstance to see in Singapore and in Borneo, among the foliage of small trees, a number of the leaves, sometimes green, sometimes brown, gathered together into a huge ball as big as one's head, about which under ordinary circumstances no ants are visible; but a smart blow upon the fabric is immediately followed by the appearance of swarms of brown ants of a large size, which soon cover the nest, and run up and down the branches in busy and terrified streams. These ants are armed with nippers, which inflict a disagreeable and startling pinch; and it is desirable not to remain long under the tree after they are disturbed. The effect of their bites, however, is perfectly transitory.

In houses ants are everywhere great pests. A small reddish species, extremely fond of sugar and other sweets, and a slender black one, both abound, and can only be kept out of the meat-safes and sugar-basins by the stratagem of immersing the legs of the tables supporting them in cups of water. This, however, does not entirely prevent their approach without further care; for a film of dust settles upon the surface of the water in the course of a day or two, forming a sufficiently stable bridge to enable the little creatures to cross over. To the insect-collector they are a terrible nuisance, for the freshly-killed butterflies, &c., are liable to be attacked and ruined in a very short time, if the ants by any accident obtain access to them. Thus, on more than one occasion, I have laid my newly-captured specimens upon the protected table, fondly believing them to be secure, when lo! after a few hours, I have found every paper swarming, and already the wings alone of some specimens left. The circumstance had arisen from the simple accident of the end of a strap lying upon the table having fallen to the ground, thus forming a convenient means of communication, of which the hungry ants had not scrupled to avail themselves. In such cases they always attack and destroy the last captured and most succulent insects. Camphor, however, is an effectual protection against these marauders, and the remedy is therefore tolerably easy if an ordinary amount of care is used.

CHAPTER XVIII.

MANILLA.

Appearance of the City—Manilla Bay—The Town—Chinese Shops—Aspect of the Mestizas—Dilapidated Condition of City—The Great Earthquake of 1863—Features of the Shocks—Their Effects—Moral Effect on the People—Game-Cocks—The River Pasig—Tobacco Manufacture—Taxes on Commerce—Sea Snakes—Tropical Skies compared with Northern—The Southern Cross—Effects of Clear Atmosphere—Moon-blindness—Case.

It was Christmas-day when we anchored in Manilla Bay—dull, wet, and dreary; but warm withal, with nothing to remind us of the season. The city looked forlorn enough, for at the best of times there is nothing very striking in its appearance, which is pretty much that of a dull continental town built in a hollow; the houses like so many barns, and the few public edifices which rise above the general level of the housetops being constructed of a dark red stone, which gives them a sombre air which even a nearer approach does not tend to remove. But when, on the bright sunny days which succeeded, the distant mountains of Luzon appeared with their changing lights and shades, forming a beautiful background to the landscape, there was much that was picturesque and attractive in the scene; while the placid waters of the bay with the distant mountain of Mariveles at its entrance, behind which the sun nightly disappeared, bathing it in rich gold and purple, completed a very charming panorama. Not always, however, is the bay so calm, for it is so extensive, that though it may be compared to a large lake, its waters are in some seasons swept by such violent winds that ships have foundered and gone down in them; and the dilapidated condition of the massive stone pier which forms the right-hand side of the harbour attests the power of the waves in displacing the huge blocks from their cemented bed.

The best built part of the town of Manilla is contained within the walls of the citadel, which is duly fortified. Here the streets are narrow and regular and tolerably well paved, the windows universally glazed with the shells of the Chinese window-oyster (Placuna placenta) in default of glass, which is very rarely seen. There are few or no shops in this part; but surrounding the citadel are the suburbs, or Pueblos, containing by far the busiest and most lively streets, with numerous good shops. The greater number and the best of these are kept by Chinese, who form a large proportion of the population, and appear to be industrious and tolerably clean. Their streets have a very cheerful appearance—a sun-awning of blue and white running along the tops of the shops, and crowds of respectably-dressed Chinese standing or sitting at doors, smoking their pipes and chatting—while the shops themselves have wares exposed in them of a far superior class to those which one is accustomed to see in Chinese shops elsewhere. One circumstance strikes the visitor as remarkably strange and anomalous—viz., the profusion of pictures of Roman Catholic saints, and prints of a religious character which adorn them; crucifixes, and rosaries, and other paraphernalia of the dominant religion, which Chinese scruples do not prevent their turning to account; and if one might judge by their abundance and prominence, it may be presumed that the enterprising Chinese tradesmen find these objects among their most marketable and profitable commodities..

The inhabitants of Manilla are said to number 300,000 natives, Spanish, and Chinese. The common costume of the men is a pair of trowsers of light material, and a kind of shirt, thrown on loosely in the manner of a smock-frock. This article of dress is most characteristic, and in it the greatest possible variety of form, colour, and material occurs. It is sometimes of linen,—white, clean, and neatly and curiously plaited and folded; but more usually the material is thin and more or less gauzy, and the colours as numerous and diversified as those of the rainbow—or in other cases black. They appear to take the greatest pride in the get-up of this article of attire, which is always clean and neat. A straw hat of various forms, more or less approaching the European, however, covers the head, and an umbrella is a constant companion, almost as constant as the cheroot. The men are very similar to Malays in aspect; but the women are very superior in this respect, being usually striking and good-looking; their eyes large and dark, and their long black hair hanging loose behind, and adding an expression of *abandon* to their luxuriant and voluptuous beauty. Their costume, gay and graceful, consists of a kind of skirt (Saya) of a bright-coloured material, and usually of large pattern, and a jacket (Piña camisa) of similar material, but somewhat scanty as to quantity, closed in front, but leaving the arms and neck bare, and allowing an inch or two of dusky skin to be visible between it and the lower garment. On their feet they usually wear high wooden sandals, which raise them two or three inches in stature, and make them appear taller than they really are. Like the men, they seldom appear without a cigarette or a cheroot in their mouths. The same dress is worn by young girls, except that their dress is often of so transparent a material that the whole form can be distinctly discerned through it; and the children of both sexes of the lower orders are often unencumbered by any clothing whatever.

These *Mestizas*, as they are termed, are the native *Indians* of the Philippines, whose blood has to a great extent probably been mingled with that of their Spanish rulers. They are a very exclusive people, speaking a language of their own, called *Tagalan;* and have their own places of amusement and entertainment, in the form of a theatre, in which the performances are of course all in Tagalan, and Mestiza balls, to which no one is admitted who does not don the costume of the country as described above. They do not, however, bear a very high character for morality,— in fact, Manilla in this respect is undoubtedly at a very low ebb.

Being a Spanish town—and a Catholic withal—the incessant beating of drums, and clanging of trumpets, is fully accounted for; and if proficiency upon these warlike instruments makes a great nation, then must the Spaniards be reckoned in the first rank. And if the jangling of bells makes a people religious, then must Manilla be a saintly spot; but one would imagine that three more discordant instruments (for the bells never ring a peal) could not have been invented to vex the ears of the inhabitants at all hours of the day and night. Two other features of the place which must strike the visitor may be alluded to—viz., the frequency of cassocked priests, not uncommonly to be seen with a cigar in their mouths; and the convicts, who, chained together in pairs, work thus side by side, and are allowed to go about without immediate supervision.

A visitor at Manilla cannot fail to be struck with the dilapidated condition of some parts of the city. Houses cracked and partially unroofed, others windowless and deserted, walls broken down, and court-yards grass-grown and uneven, piled-up heaps of hewn stones which have once been part of a building, meet the eye in every direction, and are all witnesses of the disastrous earthquake which took place here three or four years back, a repetition of the catastrophe of 1645, and which, besides destroying a great part of the city, proved fatal to a large number of the inhabitants. But the neighbourhood of the principal churches, and of the cathedral, most conspicuously testifies to the violence of its effects. These large buildings are almost totally destroyed, and are all in a ruinous condition. The cathedral has a most desolate aspect, and only a small portion remains in a sufficiently stable condition to allow of being patched up, and serving as a temporary church. Another spacious church, close by, was undergoing some attempts at repair, and huge beams of wood were in course of elevation to support a roof; but the whole aspect of affairs is melancholy in the extreme. Few attempts appear to have been made to renovate the city, and as few even to remove the débris, and with the exception of piling up the stones by the road-side, no efforts have been made to clear away the traces of the catastrophe.

The terrible earthquake which brought this destruction upon the city of Manilla took place on July 3rd, 1863, at half-past seven in the evening. Like most of these frightful occurrences, which are at the same time overwhelmingly destructive, the ruin was all completed in less than a single minute. Not, however, that this was the only shock experienced, but the only one which effected serious mischief, from the unfortunate circumstance that the city was built just upon that patch of earth which experienced the greatest throe. There are two considerable volcanos in the neighbourhood of Manilla, those of Tayal and Abbay; and although it might be imagined *à priori* that the terrific shock was in some way connected with the closing up of these natural vents, it does not appear that this was the case, for the volcano of Tayal, in the province of Batangan, was reported to have been very active at this juncture. No particular warning was given, however, of the fatal moment—it was the rainy season, and there had been at the same time much sultry weather, accompanied by heavy thunderstorms; and one of them is described as having been an uninterrupted blaze of several hours' duration, such as I have witnessed more than once, but which can hardly be considered a precursor of earthquake. In the evening, just as, it being dark, people were enjoying their cigar and the coolness of the air in the verandah, the earth shook so that they were obliged to support themselves by some object to prevent themselves from falling. Two distinct shocks immediately succeeded one another. The first was an earth-wave from north to south, which, although itself severe, would not have accomplished the destruction of the city had it not been instantly followed by another cross wave from east to west. Then the buildings fell in all directions, burying hundreds beneath the ruins. The cathedral roof is said to have opened wide with the first shock; but seemed, as the wave passed by, to subside into its original position and close up again, but the transverse wave immediately brought it all crashing to the ground. The other churches also suffered frightfully; and it most unhappily happened that it being the hour of vespers, the churches, of all places the most unsafe, were more or less occupied by people. Priests and people alike were buried in the ruins, many, of course, killed; others only maimed, but living, and their voices could be heard amidst the stones and beams which covered the floor. Energetic efforts were made to relieve them, and water was conducted through the pipes of the broken organ; but by degrees the voices ceased, and they were dead. A fine stone bridge across the Pasig was so damaged that it was deemed unsafe to cross it, and it was closed, and still remains in a dilapidated condition.

Of course innumerable houses fell to the ground, and even now many of them remain in nearly the same state as they were left by the shock—unroofed, cracked, and fissured. In one house which I visited I was assured that so great was the oscillation that the chandelier in the dining-room, hanging six feet down from the ceiling, swung so violently as to knock the ceiling on either side. A very fortunate circumstance was that at the hour at which the earthquake occurred the European population had just finished dinner, and had for the most part retired from the dining-room to the less dangerous verandah. In many places in the town fissures opened in the ground, which in some cases closed again. In addition to the immense loss of private and public property, the Government exchequer was seriously threatened by the partial destruction and unroofing of the tobacco stores. In these warehouses no less than 57,000 quintals of tobacco were deposited, representing a value of two millions of dollars; and inasmuch as the disaster occurred during the rainy season, this vast quantity of tobacco would all have been partially or entirely ruined before precautions could possibly have been taken to protect it, had it not singularly happened that the event was succeeded by a week of unseasonably fine weather. This providential occurrence was also, of course, of the greatest service in innumerable ways to the suffering population, and gave opportunity of making provision for immediate shelter and protection.

Those who were on shipboard in the bay thought they saw a phosphoric luminosity over the city at the time of the occurrence, though whether this was not conjured up by their own vivid imagination admits of doubt. More probable is the story that at the moment of the shock they felt as though their ships had struck upon a rock, a circumstance often recorded in similar catastrophes.

The city of Manilla itself seems by a curious fatality to have been the very centre of the oscillation, and not only was every pile of buildings therein shaken to its very foundations, but people who were in the town were thrown off their feet by the violence of the shock; while those outside hardly knew that anything unusual had happened. Those driving in the Calzada state that they scarcely felt any movement; and great was their consternation and astonishment, on arriving at their homes to find them in ruins, and their friends wounded and dying. It is perhaps less remarkable that persons in closely contiguous spots in the town felt the shock in very various degrees of intensity—some having been sensible of but little movement, while others, perhaps, in their terror magnified the effects which they personally experienced. For some time afterwards slight shocks were felt nearly every week; but no great and destructive oscillation has taken place since that memorable day.

It is melancholy to contemplate the position of a community such as that of Manilla, which has grown to a certain degree prosperous and important, and has raised public edifices of an imposing character at very considerable expense; but which in a moment finds itself, by a convulsion of nature, suddenly paralysed and laid prostrate—crowds of its busy inhabitants hurried to sudden destruction, their houses toppling down, and the churches and public buildings, the pride of their city, reduced to a mere shapeless mass of ruins. Still more hopeless and distressing must be the feeling that, repair and renovate howsoever they may, safety and security have departed for ever—they know not the moment when the earth may open and swallow them up in a more wide-spread and general destruction. The time may be near, or it may be far off; but it is so far inevitable that though a false security may lull the inhabitants into forgetfulness of the past, it can never inspire them with energy, or give them confidence in the future.

But the people of Manilla seem lighthearted enough, and the streets are thronged and busy. Smart carriages and pairs clatter along in the evening, full of gay occupants bent on enjoying a drive in the Calzada, which extends three miles along the beach, and which is crowded on fine evenings, particularly on band nights, when they all alight, and for two or three hours walk upon a well-lighted, spacious, and elevated promenade, listening to the strains of military music—the ladies with fans and mantillas, after the approved Spanish fashion, and the gentlemen universally smoking the native cheroot.

It is very amusing to see the passion which exists among the Manilla people for cock-fighting. I will not say that half the population go about with a game-cock under their arm; but it is a most common occurrence to see a man thus burdened, or accompanied. The cocks are very handsome birds, often of very pure breed, and seem quite at home, in town or country, tucked under the arm of their masters (who usually also have a string attached to them), from whence they look abroad complacently, and apparently in search of some other cock with which they may be permitted to fight. Occasionally two cocks thus meeting are placed upon the ground and allowed to have a little spar with one another. But restraints are placed upon cock-fighting by the Government, and it is only under licence that they are allowed to make war to the knife upon one another; but in the regular cock-fighting establishments great excitement and high gambling are often the order of the day. In unlicensed places the indulgence in their favourite sport is punishable; and were it not so the whole population would, I believe, practise it in every street of the town.

The river Pasig flows out of a considerable lake, situated at no great distance from the city, which it divides into two parts, connected by several bridges, the best of which was destroyed by the earthquake. A long mole on either side converts the entrance of this river into a harbour for small vessels—all ships of a larger burthen being obliged to anchor out in the roads, and those with considerable draught, a long way out. The tide flows in and out of this harbour with great strength and rapidity, and at the ebb always carries out vast quantities of water-cabbage (Pistia stratiotes), which is brought down from the lake. Into some parts of the town the river penetrates and ramifies into innumerable canals, among which it is easy for a stranger to lose his way—as I did, in searching for the residence of a friend. Large quantities of produce from the interior are brought down the river in barges, which are poled against the stream with an amount of labour I never saw human beings exert before. Placing the rounded end of the pole in the hollow of the clavicle, the men crawl from end to end of the barge on all fours, the pole and their body forming an almost continuous line; and particularly in passing the bridges, where I have watched them perform this painful operation for half an hour, without making any perceptible progress.

The cultivation of the interior is in a great measure rice, which is the main support of the population; but indigo is also largely cultivated and exported; and among fruit-trees, the mango (Mangifera indica) is the one which has acquired chief reputation at Manilla, where they are to be had in perfection between November and June, and are preferred by most consumers of that fruit, to those grown elsewhere. But one of the most important objects of cultivation is tobacco, the manufacture of which is taken in hand by the Government. In the factory at which the Manilla cigars are manufactured, it is said that no less than 7000 girls are employed, and the number of cigars turned out must be enormous. There is also another factory at Cavite, on the south side of the bay. But the world at large does not benefit in proportion; and so enamoured are the people of Manilla with this much-abused weed, that not more than one-seventh part leaves the island for exportation, the remainder being consumed by the population. And this can be believed, when the universality of the custom of smoking is observed in Manilla;—for not only is a cheroot the never-failing companion of the men of all ranks, but the ladies indulge equally in the reprehensible practice, and little girls even may be seen with cigars in their mouths— not of the "Queen's," or lady's pattern, but such as a professed smoker in this country would by no means despise.

The proverbial jealousy and intolerance of the Spanish

nation are well illustrated at Manilla in many ways. No Protestant church exists, or is allowed to exist there, and many English residents came on Sunday to our ship for the purpose of hearing Divine Service performed by the Bishop of Labuan, who happened to be with us. The resources of the country are cramped by the short-sightedness of the Government; and foreign trade is virtually driven away from the place by the severe exactions and vexatious imposts which are levied upon shipping. All merchant ships are mulcted in heavy port-dues, which are demanded according to tonnage; and not content with this, they add 25 per cent. to English measurement, thus materially increasing the otherwise large expenses. An English ship, driven in by stress of weather, in a partially disabled condition, just before our arrival, on completion of her repairs was not allowed to leave port until a sum of between three and four hundred dollars had been paid; while all transactions are carried on with such unpleasantness, that it is a wonder that any ships go there at all.

Lying in Manilla Bay, it was not unusual to see water-snakes (Hydridæ), swimming on the surface of the water. This family of sea-serpents is for the most part distributed in the Indian seas, though some are found about Australia and the American coast, and rarely in the Pacific. I noticed them here, and in crossing the China Sea, as well as about the coasts of Borneo and Johore. The usual appearance of these snakes is more or less variegated or striped with transverse black and yellow bars—though some are of a more uniform dark colour; and they are generally about two feet, or two feet six inches long. In calm weather, they may often be seen lying lazily upon the water apparently asleep, and basking in the sun; and they will remain undisturbed while the whole ship's length passes within a fathom of them; but sometimes, taking alarm, they will flounder about for a moment, and then dive down out of sight. When thus surprised, it is not difficult to take them in a net, for they turn over before diving—it is supposed, to expel the air, without which operation they cannot sink. But if taken, they must be handled with caution, for they are nearly all venomous, and are often much dreaded—and not without cause—since they have an unpleasant habit of crawling up the chains and through the hawse-holes, and thus getting on board ship, where they are anything but welcome visitors. They will creep about the deck—and, although I do not know an instance of any one having been bitten by them— I have known them cause considerable alarm, by getting down into the cabin, and there making their presence first known by twining round the leg of its occupant. The seasnakes, in nearly all species, have flattened compressed tails, which enable them to swim with great facility—the compression often including a considerable portion of the body. Their eyes are usually small, and the nostrils operculated or valvular. Among other Hydridæ taken in Manilla Bay, I obtained some specimens of Chersydrus granulatus —a non-venomous species, which indeed differs from others of this family in being an inhabitant of rivers, from which they are occasionally drifted out to sea. Sharks also are not uncommon in the bay; and either sharks or venomous serpents were alone sufficient to deter us from taking a delightful bath, which otherwise we should much have enjoyed in this warm place.

For although Christmas time, the weather, after the day of our arrival, was truly delightful, but little agreeing with our preconceived ideas of the season. The days were of that charming character which in England we should call perfect summer weather, but which is very rare even in the height of summer in our climate. The thermometer stood at 82° Fahr. in the shade, and of course the sun was intensely hot; but at night there was usually an off-shore breeze which kept the air pleasantly cool. And when the sun sank in purple and gold behind Mariveles, and the stars shone down in all the brilliancy of a tropical night, the scene was often indescribably beautiful. The aspect of the sky was, of course, quite different from that seen in our latitude—the Great Bear and the Pole-star having given place to the Southern Cross and the Magellanic Clouds, the wonderful Nebula in Argo, and their accompanying clusters. Much has been said about this Southern Cross, and most travellers have spoken rapturously of the glories of that constellation. That it is an interesting and beautiful one is undeniable—but one always feels how much more beautiful it would be were it a perfect cross, instead of the one-sided affair it really represents—and if δ Crucis were a star of equal magnitude with the other three. The beauty of the Southern Cross is really derived from its association with other constellations, and mainly to those two magnificent stars of the Centaur, which seem to point up to it. The Milky-Way is here, too, of remarkable brilliancy, heightened rather than impaired by the two mysterious black starless patches which show out blacker and darker the more brilliant the night. But in reality the Northern sky is nothing inferior to the Southern, so far as regards richness in constellations. Our Ursa Major has no match in the Southern hemisphere; and aided by Arcturus and Capella, Vega and Altair, the North is well able to compete with anything the South has to show;—while the incomparable Sirius, and its ally, Orion, are common to both latitudes.

It is not so much, however, the *southern* sky as the *tropical* sky which is so striking to one coming from northern latitudes. In any high latitude, the density and irregularity of the atmospheric strata produce those rapid changes in refraction which cause the twinkling so conspicuous among the stars of our own sky—a phenomenon not without its own character of beauty, owing to the brilliant and changing colours which accompany each successive change of refraction. But as we approach the tropics, the stillness and clearness of the air produce this result in a less and less degree, so that under favourable conditions the uniformity and purity of the atmosphere transmit the light of the stars with little sensible disturbance, and hence this beautiful diamond-like scintillation is more or less lost, and gives place to a placid and calm starlight, in which each orb seems to shine with the steady light of a planet, and another element of beauty is substituted for that which has been lost.

But when the full moon comes upon the scene and extinguishes the lesser stars in its effulgent rays, the tropical night is a sight to be remembered; and especially at sea, when the long track from the ship to the horizon is bathed in bright, dancing light—not dazzling, like the sun—but white and silvery, and such as mortal eyes can look upon without blinking.

The close oppressive air between the decks on such nights often encourages the sailor to carry his mattress into the open air, and sleep under the canopy of the sky—a proceeding not altogether without danger if no awning be spread, inasmuch as heavy dews often fall, and rheumatic affections are liable to ensue. If the moon be shining with its accustomed brilliancy in a cloudless sky, another danger is encountered, concerning which, however, there is a certain difference of opinion. The ill effects of the direct rays of the moon upon sleeping persons are very generally recognised among nautical men, although of course very considerable allowance must be made for prejudice as well as for superstition, and no story should be received without careful examination, and the most searching investigation, in order to exclude all sources of error. There can be no doubt whatever that thousands of persons *do* sleep in the moonlight without experiencing any ill effects, but though that fact may be admitted, it does not follow that everyone is therefore exempt. Whatever the real cause may be, it appears that young people, under 18 or 20, are most liable to suffer; and naturally, as it is impossible to estimate the predisposing influence which various shades of constitution may imply, so also it is equally difficult to ascertain what external circumstances may be most provocative of the evil believed to result. All the cases, however, that I have been able to collect have been those of lads about the age mentioned above—and when a great many such lads are on board, cases are proportionately frequent, though most generally the inconvenience experienced is but temporary and slight, and is usually best combated by the administration of tonics.

The most remarkable instance which I have been able to meet with occurred in a ship with whose *personnel* I was well acquainted; and my enquiries, made directly of those who were personally cognizant of the occurrence, elicited the following particulars, which are not without interest; and unless I was intentionally deceived, which I have not the slightest reason to suspect, they go far to prove the reality of moon-blindness.

In this case the lad was 18 years of age, of fair complexion, full face, and large, light, greyish-blue eyes, which attracted attention from their remarkable appearance. His hair and eyelashes were darker however than the colour of his eyes would lead one to expect. In February, 1864, on a certain night about the time of full-moon, this lad was sleeping on the forecastle with his face turned upward, fully exposed to the direct rays of the moon. The circumstance was remarked by his messmates, who remonstrated with him, and assured him that he would feel bad effects from it; but in spite of these remonstrances he persisted in keeping his place. Nothing occurred that night, but on the following night he was one of a deep-sea sounding party, and was beating the line, when the moon rose, and as it did so he suddenly exclaimed that he could not see, and would have fallen overboard if he had not been stopped as he was deliberately walking into the sea. For ten nights after this occurrence, as soon as the moon rose above the horizon, he complained that a cloud seemed to develop itself before his eyes, and he forthwith became temporarily blind, so that it became necessary to lead him about the deck; but this only happened during moonlight. On two occasions he narrowly escaped serious accidents from falling down a hatchway, and it became necessary to place him upon the sick-list. The surgeon, a gentleman of superior attainments, with whom I am acquainted, examined his eyes minutely, but could detect nothing abnormal in them. When the man was between decks, and out of the moonlight, he had no difficulty in distinguishing objects; nor was his vision affected during daylight, nor after dark before the moon rose. Ultimately when the next moon came round he had recovered from this singular nyctalopic affection, which did not return again.

In this curious instance, the particulars were corroborated by the evidence of all the officers and men, and the only source of fallacy is the possibility of the man having malingered. But such an idea had no apparent justification, and was unsupported by any circumstances. He was a well-conducted lad, and the fact that he was placed in serious jeopardy on two or three occasions, owing to his blindness, seems strongly to negative such a supposition.

Many other instances have been related to me by persons, sometimes medical officers, under whose direct notice they fell; and although some old surgeons doggedly refuse to give credence to any of them, and condemn them wholesale as malingering cheats, I think such a course, to say the least, unphilosophic in the extreme.

CHAPTER XIX.

HONG KONG.—CHINESE NEW YEAR, ETC.

Chinese Pyrotechny — Salutations by Crackers — Religious Ceremonies — Holiday-making — Family Groups — Children — Visits of Ceremony — Boats — Toy-makers — Mandarin Processions in Canton — Irruption of Beggars — Chinese Tame Birds — Shantung Lark — Tumblers — Canaries — Mina — Street Robbery in Hong Kong — Insecurity of the Person — Police Regulations — Contrast with Canton — Character of the Chinese — Facility of Escape to Canton.

THE 5th February (1867) was the Chinese new year, a festival held in particular honour among this people. Preparations for the day had been visible for some time before, and its advent was the common topic of conversation. It was ushered in by a great noise of crackers, which made night hideous, and rendered it very difficult to sleep, a bad preparation for the enjoyment of a festivity. The Chinese have somehow gained the reputation of being great pyrotechnists; but the display of this occasion gave me but a poor impression of their powers in this direction, which seemed to have been entirely concentrated in one channel, namely, in the construction of crackers. In this department they have certainly arrived at great perfection. These crackers are usually of small size; but great numbers are fastened together upon a string in such a manner that, when ignited, the whole series, of many hundreds, explode in regular succession with a sharp noise, like the fire of an irregular volley of musketry, which lasts ten minutes or a quarter of an hour. The bundle of crackers is suspended from a pole out of a window, and burns from below upwards; and as soon as one string shows signs of dying out another is loaded, so that the rattle goes on for an hour, or longer if caprice desire it. Another mode, of which they are very fond, is that of packing the crackers in a paper parcel, which is lighted at one corner and thrown into the street, when they explode like the firework known as jack-in-the-box. I have seen dozens of these packets thrown one after another by an invisible hand into a back yard, where they sputtered and smoked otherwise unseen, *noise* being all that is desired. It is the Chinese mode of expressing joy; and it is also an expression of congratulation when a wedding takes place, or a birthday; and it is the common custom in Hong Kong to burn crackers when a European is leaving the colony, either temporarily or permanently, in which case the members of the household make a demonstration in front of the house as the traveller quits it. So, also, whenever a ship leaves the harbour homeward bound, the bumboat alongside *chin-chins* with abundance of crackers and smoke, thus expressing their acknowledgment for past favours, and their good wishes for a prosperous voyage.

The Chinese new year is a universal holiday. Not only are all the shops and places of business entirely closed that day, but for a week or ten days, or even a fortnight, business is more or less suspended, each one taking as long a holiday as his means will allow; and during this time they superstitiously refuse to do any business, even on advantageous terms. Every house was decorated with little rectangular pieces of perforated gilt paper over the door, and a little niche in the entrance was similarly adorned and lighted with small tapers, thus fulfilling its purpose of a little shrine. The temples, or joss-houses, were crowded with devotees, who eagerly tried their fortune at the *lucky stones*, which are considered to be more than usually significant upon this occasion; and the smoke of joss-sticks and little tapers, which rivalled the atmosphere of a catholic village church on a saint's day, rendered it at first somewhat difficult to see what was going on. Men there were, and women, making the *ko-tou*, or obeisance, before the gilded idol, investing minute sums of money in paper dollars and joss-papers, which latter—squares of thin paper with a daub of gilt upon them—they took in large numbers, and having set fire to them, held them till they were in a somewhat ·dangerous blaze, and then deposited them in braziers to consume to ashes.

Out of doors the scene was peculiar, and exhibited the characteristics of the Chinese enjoying themselves in their own way. From an early hour the streets (the shops being all closed) were crowded with people walking in an orderly manner, seeing and to be seen—all well-dressed, and either exhibiting themselves or gazing at the passers-by—each one looking for an acquaintance to whom he might wish the compliments of the season, which they interchange with alacrity; the words "*koong-haye, koong-haye*" being heard on all sides, accompanied with folding of the hands and polite bows of various degrees of depth, according to the relative ranks of the individuals. Most of these parties were bound on visits of ceremony to their acquaintance, who remained in their decorated apartments and received visitors. In this case those who remained at home, as some evidently must do, performed their visits by proxy; and in all directions might be seen well-dressed servants or clerks running about with packets of red-paper cards in their hands, which they left at the houses of their masters' acquaintance with a complimentary message.

Every one makes a point of being dressed in his best on this day; and the man must be poor indeed who cannot raise, for this occasion only, a passable costume—usually a long coat, reaching down to the heels—even if he leaves it in pawn for the rest of the year. The barbers are in great request immediately previous to the festival, for every one to-day is clean shaven. The great majority of the people in the streets are family groups; and the greatest pride appears to be taken in decking out the children, more particularly the little girls, in the most gay and often the most grotesque manner. The children of both sexes are rigged out in the brightest colours, the little caps of the boys being miracles of kaleidoscopic brilliancy; while the girls' elaborate dresses, ornamented and embroidered in scarlet, yellow, and other striking colours, attract general attention. These little dolls are usually perched upon "golden lilies," encased in pretty little embroidered shoes; and their head-dresses were most carefully attended to, the hair well oiled, and brushed from the middle into tightly plaited knots on either side of the head, in which are twined gaily coloured flowers. Not unfrequently a fillet was tied round the temples, from which descended a deep fringe hanging half over the face. The child's features also had not escaped decoration—the eyebrows pencilled, and the cheeks rouged as highly as though the unfortunate was suffering from a severe attack of scarlet fever. Sometimes the rouge was nicely tinted on all over the cheeks, and had a roguish, coquettish look; but not unfrequently want of skill, or of care, had been fain to rest satisfied with a mere shapeless red daub on either side, which by no means added anything desirable to the otherwise pretty features of the child.

In company with a gentleman long engaged in business in Hong Kong, I went on a round of visits to the better class of the Chinese community, his correspondents. At every place we found a little room fitted up in a tasteful manner with pictures, flowers, and ornaments; around the walls settees were arranged with tea-poys between; a little extempore shrine, with its joss-stick taper and gilt paper; and on a little table before it, a dish, divided into several compartments, and containing a variety of assorted fruits and sweetmeats. The inevitable tea-equipage was of course everywhere; and we were invited to partake of this slight complimentary repast at each house. Every visitor who entered, folding his hands and bowing, repeated the salutation, at the same time presenting his red-paper card. If he were of equal or superior rank to the host, he was invited to be seated and take a cup of tea; but if of inferior rank, or a younger person, his visit was usually brief, and sitting down was dispensed with. Our visit was in most cases, either through Chinese politeness, or real appreciation, received with great *empressement*, and the best of tea and of sweetmeats were pressed upon us with apparent cordiality. In one house our host placed before us tea which he avowed was sold at the rate of 45 dollars, or about 10 guineas, the pound; one of those fancy articles for which the rich gave nominal and extravagant prices. He took it from a small sample canister, and it was made in the usual way—that is, by pouring boiling water upon the leaves in a covered cup, from which the infusion was drunk without milk or sugar. Not being a professional tea-taster, however, I was unable to detect the immense superiority of this tea over the more homely, but more moderately priced article to which I had been accustomed.

During the days following the new year I was at Whampoa, and also at Canton. In all places the same observances were visible; the boats were all decorated with pieces of gilt-paper hanging over the stern, while inside were small pictures of idols; and there was none that had not its little shrine fitted up, with a taper burning, and sweetmeats placed before it. Everywhere the shops were closed, and the people were parading in their best dresses; everywhere crackers exploded at intervals, and pleasure superseded business. At Canton for several days large house-boats, gay with flags, and freighted with be-rouged ladies and long-nailed gentlemen, floated into town along the "Pearl" river, mid beating of gongs and firing of guns, on their return from a holiday excursion in honour of the New-Year. Inside the city but few shops were opened until a week after new-year's day, but the narrow streets were alive with people in holiday costume. In some parts the toy-makers were doing a thriving trade, for, as with us, the hearts of the juvenile population are at this time made glad with presents of surprising playthings, often assuming the form of a lantern; but which a stranger would never suspect of being intended for that purpose. It might be a large globular fish with gay colours and expansive fins, or a gigantic frog, or a crab with moveable claws and goggle eyes, or some other nondescript animal, which is carried aloft at the end of a long stick, or suspended with a candle burning inside. Every possible variety of dolls, and of ingenious toys of the gaudiest and cheapest description, in some places almost blocked up the narrow pathway allotted to passengers. Boys paraded the streets with a flexible paper dragon, borne upon poles, the head truculent and frightful, and a serpentine movement was given to the beast by each boy waving from side to side the pole which he carried. From time to time the beating of a gong warns the passengers to stand aside as well as they can, to make room for a mandarin who is going out on a ceremonial visit. And a shabby procession it is; for no mandarin is seen abroad without his retinue, though they appear to be in no wise particular as to the character or appearance of its elements. Some dirty and scantily-clothed boys, carrying a gaudy flag or two, follow the man who heads the procession beating a gong to clear the way; then comes a man bearing a gigantic fan, followed by one or two spotted or piebald horses, with an attendant at the bridle, and after them is the great man himself, his sedan borne on the shoulders of four or six men according to his rank, from out of which, with his hands folded upon his portly person, he looks impassively and sleepily through his great round spectacles, a momentary glance of something like interest falling upon the western foreigner who is standing aside (perforce) to make way for him. The peacock's feather, and button in his cap, and the embroidered bird on his breast, are the marks of his nobility, and precious to him as the means whereby he squeezes out of his dependents an income at least quadruple that allowed him by the law. One or two more horsemen form his rear-guard, and the procession is closed by a couple of men carrying an old portmanteau or bandbox on a bamboo across their shoulders; but whether this contained a change of costume, or presents, I know not, but only that it always forms an integral portion of the mandarin's train.

Another feature of the New Year in Canton is the irruption of beggars. Being a time of year when everyone is anxious to *raise the wind*, the beggars seize the opportunity of the general good-humour and festivity to endeavour to do the same by appealing to the compassion, or if that fails, to the risibility of their countrymen. A ragged beggar here is a sight, for there can be no doubt about his rags; but one feels a difficulty in accounting for the manner in which such a heterogeneous mass of tatters is held together. Their filth, too, is extreme, and they swarm with vermin. Some have their faces painted like a clown in a pantomime, and make grimaces and attempts at jokes, which I, for one, could not appreciate; others were dressed in women's clothes, and smirked and talked *falsetto* to the amusement of the passers by; some carried a monkey, like an Italian organ-grinder, and sung a Chinese ditty; while others absolutely howled and writhed about as if they were suffering agonies; but as the dodge was well known and understood, no very lively sympathy was exhibited, though they probably earned their proportion of cash. But the most common method of exciting compassion was to go from door to door with hair unkempt, and dirty rags hanging from a dirtier person; while streams of clotted blood trickled down the face as though from a gash in the forehead. But I looked at several such objects narrowly, and became convinced that no such gash existed; but that the butcher's shop had afforded the gore, which was innocent of ever having flowed in any other veins than those of a pig or a sheep.

Such are some of the scenes which inaugurate the Chinese new year, in which much more character is seen than during the hum-drum round of every-day life. They have many other festivals, and are fond of holidays; and each festival has its special characteristics, but this one I had the best opportunity of observing in the streets of the great southern capital.

The Chinese are very fond of keeping tame birds; and it is a common sight to see a Chinaman leisurely walking the streets with his bird-cage, usually round and arched, upon the open palm of his hand, the wrist being bent back and the palm upward. He loves thus to give his bird an airing, as well as to exhibit his treasure, which is not unfrequently of considerable value. The bird thus favoured is, in nine cases out of ten, the Shantung lark (Acridotheres cristatellus), which is not however a true lark, but a starling—a pretty bird, nearly as large as a thrush, of a mottled-brown colour, with a light streak over the eyes, and an irregular black ring round the neck. The eyes are small, black, and have a remarkably pleasant look, and the cheeks swell out below the eyes in a peculiar manner. The natural habits of this bird are characterised by familiarity; and they have received the name of Pako, or the eight brothers, from the Chinese, because they are usually seen in small parties together. The bird is lively, good-natured, and easily tamed; but it is none of these qualities which specially endear it to the Chinese, though they all add to its attractions. It is its powers of mimicry which render the Shantung lark so popular. They have, it is said, a good natural song, not unlike that of a skylark, which I should doubt; but they easily learn to imitate all manner of out-of-the-way sounds. They will bark like a dog, mew like a cat, crow like a cock, or cough and sneeze like a human being. Nor are powers of speech denied to them, for they learn to talk with as much facility as a parrot. It is no wonder, therefore, that well-educated birds command a good price. I have known one in a bumboat for which 25 dollars (6*l.*) have been offered and refused; and I am credibly informed that good birds fetch 50 or even 100 dollars, the plumage being considered by good judges as one of their important *points*.

Although this bird is the universal favourite in China, there are in the bird-shops many other interesting species. Among these may be particularised the fork-tailed Parus (Leiothrix luteus, Scop.), a bird which, if it could be introduced to English bird-keepers, would undoubtedly prove very popular. It is a remarkably pretty bird, in form and habit strongly reminding one of the English robin, which it also equals in size, but has a stouter build. The beak is bright red, throat orange-yellow, back olive-green, tail black and forked, legs yellow, and wing primaries edged with bright yellow and deep red. The eyes are black and brilliant, and the gestures and habit of the bird lively and interesting. As only a dollar was demanded for one of these birds, including a good cage and abundance of seed, it is not wonderful that several were purchased in the hope of bringing them safely through the homeward voyage to England. Being an insectivorous bird, however, I always had strong misgivings of the result, and eventually they all died before reaching the Cape, except one which survived a few days later, and this, notwithstanding that they were fed with some half-dozen living flies nearly every day.*

But the most remarkable feature in the Leiothrix was a curious habit they had of turning somersaults on their perch.

---

* It is worthy of remark that during a passage of five or six weeks from Java-head to the Cape, there were always plenty of common flies in the wardroom. We were at no time plagued with their numbers, but several birds were daily placed upon the table, and it was always an easy matter to catch upon the walls and rafters sufficient for their delectation. The birds took the flies readily from the hand, in numbers varying from three or four to six or eight for each bird daily.

Throwing the head far back they would turn over, touching momentarily the bars of the cage in passing, and alight on their feet, either on the floor of the cage or on the perch, repeating the operation rapidly and constantly, and not unfrequently turning over in little more than their own length. When I first noticed this freak in a bird-shop I set it down as a matter of education: but I have since found that every individual has the same habit, although some tumble better than others. The Tumblers, by which name they came to be generally known, had a short, loud, and somewhat monotonous song, not unlike that of a missel-thrush, and often when placed in different parts of the ship I have heard two singing alternately in reply to one another for an hour together.

Canaries are also in plenty in the bird-shops; but Japan seems to be the paradise of the canary-bird. The "Scylla," homeward bound, was like an aviary. On a sunny afternoon I have counted 50 or 60 cages on deck, few containing less than two, and some as many as seven or eight birds, all singing in chorus. The attraction was that in Japan good singing canaries could be purchased at the rate of an itzeboo, or about one shilling and sixpence each; and the sailors, therefore, had made their hay where they found the sun shining.

The Grackle (Gracula religiosa), called in these parts the Mina, is a favourite bird, much admired in Singapore and Borneo. It is as large as a jack-daw, black, with long feet, and two yellow wattles on each side of the head. Like the Acridotheres above described, the mina is also a member of the Sturninæ, or family of starlings. The powers of imitation of the human voice possessed by this bird are truly remarkable. The Governor of Labuan possessed one which was a good specimen of its class. At my first visit to Government House, just as I reached the door, I heard a loud and perfectly distinct voice shout out, "Orderly, call the boy;" and then, "What do you want him for?" very clearly enunciated; and this was immediately succeeded by a loud laugh and a sonorous whistle. I looked in vain for the source of this unseemly exhibition; and when, presently, the same sounds proceeding from the verandah, I went out to see what they meant, the innocent-looking black bird hopping about demurely in a wicker cage would never have been suspected, had he not burst into a hoarse laugh the moment my back was turned. They will imitate a child crying in a most painfully natural manner; and their mimicry of the human voice is far superior to that of a parrot, being perfectly free from ægophony, and loud, distinct, and clear in enunciation and utterance.

Before I arrived at Hong Kong I had been told stories of persons having been attacked in broad daylight, knocked down and robbed by Chinese roughs and thieves; but while on the one hand such stories were rife, on the other I met with persons who had long resided in China, and who assured me that there was no danger of any such attack. So that I was the more ready to give credence to the latter than to the former, and the tales of highway-robbery and violence which I had heard made no impression upon my mind. I was destined, however, to be undeceived in my own person; and less than a week after I had set my foot in China I was myself the victim of one of those atrocious outrages which are but too common in Hong Kong, and are a disgrace to the Government of the colony. Feeling no sense of insecurity while surrounded by busy crowds of people, I naturally, as a newly-arrived stranger in so interesting a country, went into the streets for the purpose of making myself acquainted with the Chinese people at home, pursuing their avocations within doors and without—buying and selling, eating and drinking—all of which, and much more, may be seen as one passes through the thoroughfares, and would naturally attract the attention of an observant new-comer. I had walked down Queen's Road, the main street of the town, and, intending to make a slight *détour*, turned into a street leading up the hill. In China there is not that difference in streets that one sees in England, and it is not so easy to perceive at first, either by the dress of the people or other signs, that one street is greatly inferior to another. It was just mid-day, and the streets through which I was walking were thronged with people, either passing to and fro, or standing at the doors of their houses, or looking from their windows; but they were all, without exception, Chinese. Having gone a short distance up the street in question, I crossed into a parallel street, intending to descend into the Queen's Road again, and was so descending when I found myself suddenly in the midst of a knot of some eight or ten Chinamen. There was nothing in their dress or appearance which directed my attention to the probability that their object was robbery or outrage; and I was just passing on, when they made a simultaneous rush upon me and pushed me down, one of them striking me in the face, but so suddenly and unexpectedly that I had not a moment's opportunity for defence. While several pinioned me on the ground one unbuttoned my coat and detached my gold watch and chain, upon which they all made off, leaving me to gather myself up as I best could. Seizing my hat, which had, of course, been knocked off in the scuffle, I started instantly in pursuit, being but a few yards behind the scoundrels; but they knew their ground, and I soon saw the folly of pursuing, alone and totally unarmed, a band of Chinese thieves into their fastnesses; and seeing them all turn into a narrow slum, I retraced my steps with the intention of at once informing the police. Not fifty yards from where the robbery took place I met a Malay constable, whom I took with me to the station and saw the superintendent of police, to whom I stated my case, and gave a description of the stolen property. An inspector and a Chinese interpreter, &c., were at once despatched with me to the spot; but it was impossible for me to do more than point out the place where the affair had occurred. As for recognising any one who was standing by, every one who knows the Chinese knows also the impossibility of distinguishing one Chinaman from another, unless he is personally acquainted with one or both of them; and I was therefore unfortunately entirely unable to identify any of the numerous rogues who stood coolly looking on while the attack was being perpetrated. A number of men loitering about the spot were arrested as suspicious characters, and their tails being tied together they were carried off to the police-station; but nothing could be proved against them, except that some of them were "old offenders."

Now here surely is a circumstance calling for the gravest attention, and most vigorous correction. An Englishman walking at mid-day in the crowded streets of an English colony, having a governor, magistrates, and establishment of police, may be knocked down with impunity, and robbed in the presence of a hundred people, who coolly look on and smoke their pipes during the performance, as if it were the most ordinary and common-place occurrence. Nor is this the worst—it *is* a common occurrence for an Englishman (usually a stranger) to be so robbed, irrespective of time and place; but unfortunately it is equally common for the unhappy victim to receive a fatal stab from a knife, or a cowardly crushing blow upon the back of the head with a stone or heavy bamboo, which, rendering him insensible during the robbery, also further endangers his life.* From this complication I can never be sufficiently thankful that I escaped, perhaps because of my perfectly defenceless condition, which rendered me an easy prey to so many assailants.

There is no denying the fact, therefore, that robberies with violence are by no means uncommon in the streets and neighbourhood of Hong Kong. It is unsafe to walk in many of the streets even at noonday—it is unsafe to walk alone in the suburbs—it is unsafe to go almost anywhere after dark, without taking due precautions. A certain improvement, it must be confessed, has taken place recently. No Chinaman is permitted to perambulate the streets after dark without a proper pass, which is a partial preventive; and again, every boatman taking a passenger off to a ship at night is obliged to show his number to a policeman, who also takes a note of the ship to which he is going—a regulation which ensures a certain amount of protection in what was a few years since a hazardous proceeding—for it was

* As a good, but not uncommon example, I subjoin the following, cut from the "China Mail" of February 28th, 1867.—"As the captain of one of the vessels in the harbour was passing the Peninsular and Oriental Office, about 6 P.M. yesterday, he was suddenly struck on the head by a Chinaman with a large stone and thrown to the ground insensible. He remained in this state for some little time, being picked up by two Europeans passing at the time and carried into a Chinese shop close at hand. For some hours he was unable to do anything, considerable bleeding from the internal part of the ear continuing for a long time, and a fresh hemorrhage took place this morning. The only thing stolen from him was his umbrella, the thief being probably disturbed by the approach of the Europeans above referred to. The scoundrel has not yet been identified,"—nor ever will be.

formerly highly dangerous to entrust oneself to the tender mercies of the boatmen after dark; and even now from time to time sailors taking a boat without this precaution are robbed and thrown overboard by the miscreants into whose power they thus fall. I have known of an instance of a lady being torn from her sedan by a gang of scoundrels while passing through a quiet street, and of children stopped on their way to school; barbarous murders from time to time take place; and of late, gang-robberies with brutal violence have once more become rife.

It must not be supposed, however, that all parts of China are equally insecure as Hong Kong; for the most remarkable circumstance is that the reverse is the case, and that in almost any part of China life and property are more safe than they are in the English colony. In Canton, one may go about in perfect safety, and a lady who had resided there for many years assured me that she would have no hesitation in walking alone in any part of Canton, or at any hour. Under the Imperial government the laws protecting life and property from violence and robbery are well calculated to be effective, being enacted by those who understand the people with whom they have to deal; but in Hong Kong the legislation which is directed against the offences of Europeans, utterly fails when applied to those of a people of so entirely different a spirit as the Chinese. The Chinese are Easterns and Pagans, and they have all the faults and vices which characterise Easterns and Pagans. They have no regard for truth, but are proverbially and systematically a nation of liars, who do not know the value of truth—have no inherent love of it, and think it no harm to cheat and deceive. It is therefore the most difficult thing in the world to know how much of a Chinaman's statement may be believed. They are pagans of the lowest type, superstitious to a degree, and place the lowest possible estimate upon the value of human life. They have no fear of death themselves, and will sell their own lives for a small sum of money; so that it is not to be wondered at that they hold the lives of others cheaply, and would commit murder for a dollar, if there was a tolerably good chance of their escaping detection. The ordinary punishments, therefore, which fail to deter from crime in our own country, are still more ineffective when directed against such a people; and the leniency with which the offenders are treated is not only utterly unappreciated by them, but is mistaken for fear, or inability to act more vigorously. A Chinaman in his own country does not meet with much consideration, and perhaps the system of administering justice is not so perfect as with ourselves; but though possibly in China there is no particular dread of punishing innocent people, at the same time the measures taken are undoubtedly much more deterrent from crime than our own.

Our own administration of justice might be no less impartial than it is, while yet a difference might be made between the punishments appropriate to European Christians, and those suitable to Eastern pagans. Our prisons at Hong Kong are comfortable, and the food plentiful and good, and a sojourn in them is not feared by a Chinaman, who knows perfectly well that he is safe from personal injury when in the hands of the Westerns. Trial by jury with Western forms and ceremonies, hedged in by oaths and adjurations, is a farce when Chinese rowdies are the defendants; and however harsh this may sound to English philanthropists, it will be confirmed by the large majority of residents in China. The only deterrent used by the

Hong Kong court is the whipping-post—a means but little employed, but which has greater effect with the Chinese than fear of death itself; and were it employed more frequently—were it more inevitable than it is, crimes such as those I have described would be greatly diminished. But the Colonial government, however inclined to protect themselves in this respect, must bow to public opinion in England—the opinion of people who must judge upon abstract principles without any reference to the necessities of the case.

Another reason which has a powerful influence in regard to the statistics of crime in the colony, is due to the fact of the proximity of Canton, and the frequent and free communication which takes place between the two places. A steamer of the American type runs daily between Hong Kong and Canton, which, while it charges six dollars for European passengers, takes any Chinaman for one dollar; while the whole lower deck is devoted to the lower class of Chinese, who pay a quarter of a dollar (or 1s.) only for their passage. Of this class the steamer daily takes a crowd, who go freely to and fro with no police supervision, and thus every day a mob of rowdies comes down from Canton, where there is little exercise for their abilities, to practise upon their natural prey—the Western barbarians in Hong Kong; and having committed a robbery or a murder may return in the morning, and become lost in the world of Canton—out of the reach, and beyond the authority of the English police. It is true that if such a murder or robbery is discovered before the steamer leaves at eight in the morning, the police examine the departing passengers; but if the offender can pass muster, or can defer his departure, he still can get away; and once gone, is infinitely safer than a forger who has decamped from England to America. Latterly the police have so far overcome prejudice as to brand notorious characters upon the left ear, and see them safely off to Canton, with the understanding that their being found in Hong Kong again will be sufficient for their condemnation.

Returning for a moment to the attack upon myself, the Governor, the Admiral, and many other leading people in Hong Kong, were well acquainted with it, and I had conversation with them all upon the subject; but nothing was done which could have the slightest influence in abating the evil. Of the two Hong Kong papers, while one—the "Hong Kong Daily Press"—gave a proper and authentic statement of the circumstances, the other spoke of the matter in the most flippant style, as though it were a good joke. A distinguished naval officer, who had been several years in Hong Kong, and who told me that he had himself been several times attacked, expressed to me a strong opinion that the proper course would have been to lay an embargo upon the street in which the outrage took place, in order that the people might be taught that connivance at robbing implicated them in the crime, and that if they chose to look on unconcernedly while a person was being attacked and robbed, they must take the consequences. It is well known that this is the Chinese law; and that at Canton, for example, if a robbery took place in the street, the assailant would at once be seized by the bystanders, who are well aware that if they did not do so, but allowed the robber to escape, they would themselves be held accountable, and punished accordingly,—a wholesome law, which is much wanted in Hong Kong.

CHAPTER XX.

CANTON.

Strangeness of Canton—Bogue Forts—Whampoa—Pagodas—Approach to City—Boat Population—Pic-nic Boats—Streets of Canton—Chops—Puntinqua's Garden—Fa-tee Gardens—Gold-Fish—Deformities—Diet of Chinese—Dog-eating—Salt Monopoly—Unity of Chinese People—Its Causes—Insurrectionary Movements—Influence of Western Civilization—Benefits of Western Trade—Pekin Memorial on Western Education—Proposed Introduction of Railways—Language the Great Barrier—Prospects of Christianity.

OF all the cities of China, it appears to be agreed by travellers that none is more worthy of a visit than Canton. Doubtless there are peculiar features in Pekin, which render it specially interesting. Pekin is the royal city, and the great capital of the North, and situated so far apart, and in so different a climate from that of Canton, that it must necessarily differ greatly in character and points of interest from the latter. But Canton is the great city of the South, as its name implies,—and for strangeness, for wonders, for novelty, it is really unique. All the numberless contradictions of the Celestial Empire may be found here in small compass—all one's ideas of Chinese customs, architecture, and modes of life, imbibed from earliest infancy, here find at once their embodiment and their correction; and while, on the one hand, everything is strange and *outré*, on the other one feels a familiarity with the details of the scene which insensibly imparts an air of reality to what would otherwise appear more like a chapter in the history of Aladdin. I cannot, under these circumstances, therefore, ignore two visits I paid to this great city of the far East, although I must necessarily confine myself to the salient points which appear most interesting to the traveller, and bear as much as possible upon the main object of the present work.

The river passage between Hong Kong and Canton is not particularly interesting nor picturesque. The most noteworthy spots are the Bogue Forts, and Tiger Island, whose association with the operations of the allied armies at Canton render them historical. The whole coast, where visible from the channel, partakes of the general sterility of the Chinese shores; but it is not until we reach Whampoa that the river narrows so as to be river-like. The shores are here flat, and rather uninteresting; but the monotony of the view is broken by two nine-storeyed pagodas, built as usual upon knolls which rise somewhat above the level of the country. At Whampoa a few European ships lie in the river or in the docks, but the trade with this port is now small compared with what it formerly was. There is nothing to detain us at this dirty Chinese town, especially as we are on the way to a city of such importance as Canton. The daily steamer from Hong Kong (built on the American river-model) calls here, and in the short distance which remains, the Chinese characteristics exhibit themselves at every turn. The alluvial soil on either side is highly cultivated, and much produce is constantly diverted from the market-gardens on these banks both to Hong Kong and to Canton. Banana plantations and rice fields abound, as far as the eye can reach—the stacks of rice hanging over the water's edge, for the better facility of loading boats. Large salt-junks cluster in the neighbourhood of the salt-excise house, which is about half-way between Whampoa and Canton; and every now and then gaily-painted war-junks, with highly-decorated flags, and a great pair of eyes painted on the bows, drop down the river, their sides bristling with awkward-looking guns.

The approach to the city of Canton is not architecturally striking. The pagoda in the Consul's grounds, and the celebrated five-storeyed pagoda upon the heights, are the only salient features—if we except the great unsightly windowless structures which are used as pawnbrokers' warehouses. The river, poetically called the "Pearl," here takes a bend to the left, leaving the White Cloud Hills behind, and presenting a flat country in front, upon which the city stands. Crowds of boats, and tiers of great junks brilliantly painted, and usually ornamented with an elaborate eagle upon the broad stern, form the most remarkable features, and constitute a moving panorama of great singularity and novelty. But the houses themselves, as far as regards those lining the river, are not very unlike those which we see on the banks of the Thames, and viewing those alone, the traveller might almost fancy himself at Wapping, in the neighbourhood of the Thames Tunnel.

Nevertheless, there are few more extraordinary places than the Canton River, supporting as it does a vast population, which inhabits the numberless boats of all forms and all sizes. These boats, however, are nearly all moored, and arranged in such a manner as least to interfere with the navigation. They consequently form streets, in some parts, of a novel and striking character, full of stirring life and bustle. Outside these streets the sampan or junk plies its way unimpeded, and the *boat-wall* presented in this direction is comparatively dull and inactive, like the back of a row of houses; but direct your boat through the avenue, and all is bustle and activity. The tide is very strong, and it requires all the energies of the clever boatmen and women to make way either with or against it, through the crowded thoroughfare between the rows—a highway by no means silent, but constantly resounding with the cries and objurgations of the busy Chinese, who are now rowing, now pushing with a boat-hook, now threading their way through the craft which are moving in both directions, now bumping against the stationary boats, and thus making slow progress up the street. Every such boat has its family dwelling in it, and each presents its little scene of domestic life before the passing eye. Besides the sampans, or common covered boats, there are many palatial craft, with elaborately-carved and gilded fronts, which in the evening show a blaze of light, and busy waiters moving about among the feasting Celestials and painted Chinese women mixing with the crowd; not unfrequently gambling-houses, or places of licentiousness and debauch. It is, altogether, a scene not to be forgotten; and, as night advances, the streets of boats are extended by the crowds of sampans which have been plying during the day, but which at sunset take up their stations side by side in the canals, within which they are secured by a boom, just as the gates of the city are kept closed during the night. As evening comes on, also, numerous large house-boats, two storeys high, richly decorated and ornamented, return from their various pic-nic excursions—a number of half-naked Chinamen poling them slowly and laboriously along; meantime, groups of the better class stand at the door enjoying the scene; and others may be seen, through the windows, seated in the saloon, drinking tea and smoking, while the upper windows disclose many fair ladies in their boudoirs adorning themselves for the delectation of their lords. Trading-junks and passage-boats crowded with all classes of Chinese swell the scene; and as they arrive abreast of the town, a man standing on the high stern of each, beats vigorously upon a gong, an exercise of a religious character, which makes a din of a most unmusical and barbaric character, but which, repeated every evening, soon falls unheeded upon the ear.

The great feature of the streets of Canton, next to their narrowness and the badness of the pavement, is the wonderful variety of *chops* which hang suspended from the various parts of the houses. These chops are usually boards of all sizes, and most variously coloured, and all of them bearing Chinese inscriptions in letters of every degree of magnitude and of the brightest hues. Sometimes the letters are green on a black ground, or gold on black, or red on gold, or white on red or brown, or vice versâ, and the gaiety given by them to the scene is indescribable. In some cases these chops extend across the street, or along the front of the shop; but in most instances hang suspended and facing the passenger. They contain not only the name and business of the shopkeeper to whom they belong, but are often inscribed with some philosophical or pious sentence from the classics. The shops themselves are all open to the street, without any fronts beyond a kind of counter, upon which the wares are exposed; and the passer-by may witness in them almost every feature of Chinese industry, fan painting, ivory carving, silk weaving, toy making, idol painting, &c. But I must leave to others the task of describing in detail the lions of Canton, which have been the subject of many papers and sketches, and will provide matter for yet many more.

There are no public parks or places devoted to recreation in Canton; but certain private gardens are much frequented, especially those known as Puntinqua's, belonging to a wealthy Chinaman of that name. These gardens are extensive, and costly in the character of their decoration, containing numerous summer-houses, terraces, marble walks, fish-ponds, &c.; but all in a very dilapidated condition, as, in fact, are nearly all places in China. They seem to have no idea of keeping buildings in a state of neat repair, and the result is a very great drawback to the effect of works upon which large sums of money have been expended.

More interesting, however, in a botanical point of view, are the Fa-tee gardens on the Honam side, which are, in fact, nurseries, in which are cultivated vast numbers of plants for the supply of the private gardens of the Chinese. Here may be seen also numerous specimens of horticultural ingenuity, and dwarf plants, miniature trees and shrubs curled and bent in every imaginable form, and trained, besides, into the forms of animals and other objects—frogs, pagodas, baskets, elephants and castles, fans, stags among trees, human beings, fish, sampans, cats, scrolls, vases, &c., &c. These grotesque plants are usually dwarfed and trained over a wire framework, made of the form intended to be represented. They are kept carefully clipped, and suggestions are added to keep up the illusion, in the shape of egg-shells with a black spot to represent eyes, painted faces, feet, &c.; so that it is by no means difficult to recognise the intended shape. The dwarfing is effected in the usual way, by confining the roots in small pots; but I saw none of those wonderfully minute specimens, in which the Japanese so greatly excel. It is amusing to see the Chinese gardener water his choice plants. Taking as much water as his mouth will hold, he squirts it out all over the plant in a fine rain, as effectually as if he had performed the feat with the aid of Rimmel's patent vaporizer. I have seen them water linen in the same way in the process of washing.

In some vases containing gold-fish I observed a most singular variety, which if seen depicted would have been almost regarded as a work of imagination. Not only had they a double caudal fin, which is not an uncommon variety, but the expanse of their tails was so great that it might almost have been said to have a triple tail; while the eyes projected so far from the head as to have the effect of being seated upon veritable footstalls, and bearing a resemblance to those of the telescopic carp (Cyprinus buphthalmus). This curious variety I have seen figured in rice paper drawings, and representing so *outré* an aspect that it has condemned the whole book as one of fabulous animals—most unjustly, however, for it was a faithful representation of a not uncommon fish; and the same may be said of other rice-paper drawings, for although often highly coloured, owing to the brilliancy of their pigments and the remarkable facility for taking colour which characterises rice-paper, there is, on the whole, a considerable amount of fidelity in most cases to the objects they profess to represent. I also met with this fish in papier-mâché in the toy-shops, in form correct enough, but coloured *à discrétion*.

In China, owing to the scantiness of clothing, any deformity of course becomes very apparent; nevertheless, very few arrest the attention even when directed to the subject. Neither in Hong Kong, Canton, nor Shanghai, did I observe a hump-backed Chinaman; but in Singapore I noticed four or five, and one hump-backed Kling. Effects of bad surgery and neglect are not unfrequently seen in the form of horrid ulcers, foul wounds, and carious bones; and at Shanghai I saw an unfortunate little boy sitting at the door-step with both his feet cut off, and the bones protruding upwards of an inch through the discoloured and ulcerated skin, which was covered with flies; but such a case as this was probably a monument of the atrocities of the Tae-ping rebels, as the Chinese do not amputate surgically, and secondary amputation would much have ameliorated the poor boy's condition. On another occasion I spoke with a Chinaman whose right arm dangled uselessly at his side, wasted to mere skin and bone, and looking as though it did not belong to him. His shoulder had long since been dislocated into the axilla, and the dislocation never having been reduced, the maimed limb had wasted away for want of use, and a false joint had ultimately been established.

It can easily be imagined, in fact, that in a country where surgery is at so low an ebb as in China, the unhappy victims of accident or surgical disease are either by degrees totally unfitted for active life, or pine away and die for want of the necessary relief; and this may be the explanation of their rare occurrence as public spectacles. It is frightful to contemplate the amount of suffering entailed upon such unfortunates, whose cases, through ignorance of the correct mode of treatment, are neglected, until Nature slowly and painfully performs an imperfect cure, or the unhappy victim succumbs.

The Chinese, however, have a very proper respect for barbarian surgery; and at Canton, the hospital and dispensary, established by the American medical mission, are daily crowded with patients, who exhibit every phase of medical as well as surgical lesion, and are in large numbers skilfully treated by Dr. Kerr.

Much has been said about the diet of the Chinese, and the strange articles which occasionally enter into it. Rice is undoubtedly the staple of their food, although they often indulge in some small quantity of animal food in addition, if they can afford it. It is a very common thing to see a Chinaman carrying home his dinner or his supper in the shape of a little fish—perhaps two if they are very small; or a minute pork chop dangling at the end of a piece of grass; in either case the morsel being such that an English labourer would swallow it at a single mouthful. Pork is undoubtedly their favourite meat, and pigs are kept in great numbers, and always form an integral portion of the population of a Chinese village. They are great, ugly, hollow-backed, black animals, their bellies sweeping the ground as they wander about in search of food, as to the quality or nature of which they are not at all particular. In fact they are literally omnivorous, and no one who has watched their habits could eat them unless he were either a Chinaman, or were starving. These domestic pigs are believed to be derived from the stock of the Sus leucostymax (Temm.) of Japan. In country places the Chinese are by no means nice, eating everything that is eatable, and when by the sea-side, living, as I have elsewhere observed, on shell-fish of all kinds with little or no distinction. Like the French too they eat frogs, and in Formosa I partook of that delicacy "as in France"—the species eaten in this being Rana tigrina. I also had offered me there a freshwater turtle (Trionyx sinensis) for the larder. But in a large place like Canton, other articles are included in the bill of fare, to meet all conditions of purse where a man cannot forage for himself. Dried provisions are here very much esteemed; the small ducks which are sent out to feed in the duckboats are usually cut open and made perfectly flat and then dried; and a man will hawk about near a hundred such dried ducks strung on a pole across his shoulder. What particular delicacy there can be in ducks'-bills I did not make trial of, but they are common articles hanging suspended in the provision shops. So also are dried rats, similarly split open and hung up in front of the shops for sale—their rodent teeth betraying them in their otherwise disguised condition. But dogs are never seen in this respectable situation; nevertheless dogs are eaten in Canton, and that largely. The dog consumed by the Chinese is of a small size and usually of a light brown colour, covered with a coat of soft, short hair, so thick as to look almost like wool. But the Chinese housewife refuses to cook dogs in the family pot, or in the domestic kitchen, and they are driven to the alternative of being boiled in the streets. On any morning, in certain open spaces at street corners, the execution of a certain number of unfortunate *chow-chow* dogs may be witnessed; after which, having been skinned, they are forthwith placed in a suspended cauldron, and the *disjecta membra* are there to be seen simmering, and inviting the passer-by to stop and dine, which they do there and then.

But whatever be the nature of his diet the Chinaman consumes a large amount of salt; and salt is a commodity for which the paternal government makes him pay an exorbitant price. Salt is in China a government monopoly, upon which a large duty is payable, and no foreign salt is allowed to be imported. It may easily be calculated that two millions of tons of salt are annually consumed by the

Chinese, who pay from 10*l*. to 20*l*. per ton for it in various parts of the country, although its cost of production is somewhere about 3*l*. or 4*l*. per ton. Indeed English salt could be sold in China for 3*l*. per ton, were its importation allowed; and did the Chinese sell it at half the present rate, it might yield a revenue much greater than it does at present. But unfortunately their political economists have not learned the important principle that the reduction of a necessary article from a high price would greatly increase the demand, and that low taxes produce a revenue equal to that of high duties, by promoting consumption. It is difficult to persuade a people who run so evenly in the same groove, that a radical change in the collection of so certain a source of revenue as salt can possibly be beneficial; and the time has not yet come when foreign salt can obtain a footing in China—though indeed an approach to that desirable end may be perhaps foreseen in the recent recognition of private salt-factories in the Chusan islands.

No visitor to Canton can fail to be struck with the unity of the Chinese people, and their remarkable consolidation as a nation. The curious method of dressing the hair gives them all an extraordinary general similarity of appearance, so that it would be more easy to distinguish a Chinaman in a crowd than a man of any other nationality whatever; and the stereotyped form of their costume assists in establishing this aspect of unity. And when reference is had to the undoubted age of the Chinese Empire, and to its immense extent, no less than to its wonderful isolation, it is a moral and political phenomenon which has not its equal upon the whole globe. For no one can live long in China without becoming aware that the defects in its government are of the most frightful and glaring kind, such as not for a single year could be tolerated in the West. Extortion on the part of officials, peculation in every grade of official life, the grossest inhumanity, contempt of life, a venal justice, insecurity of property among the middle classes, and of position among the higher Mandarins, who are liable at any time to be disgraced, even though well-intentioned, by the occurrence of a mere accident—all these blots deface the Chinese system. How comes it then that the Empire has been so long and so firmly established? The inculcation of filial piety, and the habit induced by strict education through a series of generations, of giving honour not only to living parents but also to their progenitors who have long since ceased to live, except in the shrines of their surviving posterity, are doubtless most salutary, and have had something to do with the remarkable phenomenon. But this patriarchal government alone is not sufficient to account for it. The real secret lies in the system of literary examinations, and their fruits. Every man in China is aware that his talents, if duly improved, will lead him to office and power; and it is open to every one of them—not as in America to be President—but to rise to important and lucrative posts in the Empire, if they distinguish themselves in the examinations, to which all classes of Chinese periodically crowd. These examinations are conducted with such a rigid regard to impartiality, that although perhaps the only part of their system which is not rotten at the core, none but the best men carry off the highest prizes—and these are the men of talent, who are not lost sight of in the distribution of posts. None but these are advanced to important offices, or governmental departments—they are looked upon as the wise men of the land—they are the talented, to other than whom it would be folly to intrust the offices of state. This principle is so thoroughly established in the Chinese mind, and so taken in with their mother's milk, that those who do not possess the requisite amount of genius or industry look upon their more fortunate brethren without repining, being fully convinced that the principle is acted upon with impartiality—so that on the one hand all the public business is in the hands of that class of men who are most capable of performing it in a satisfactory manner, and on the other the most able men are all officially employed and well paid, and therefore the least inclined to disturb the *status quo*, while the reformer and demagogue must be drawn from the ranks of those who have either failed at the public examinations, or have not had sufficient talent or ambition to induce them to make an effort to succeed—negative qualities which are no less against their succeeding as agitators. Hence they are at a great disadvantage; and however much right and justice they have upon their side they cannot fail to be in a minority.

But although the Chinese Empire has lasted in its integrity for it may be so many thousand years, that is no reason why it should be an exception to the careers of other nations, and should last for ever. And indeed in the present generation it has sustained some severe shocks. Notwithstanding the terrible lessons they have had in the Tae-ping rebellion, the Chinese administration, as soon as they are clear of one difficulty, relapse into torpor, inactivity, and, what is still worse, oppression. Instead of keeping up a force which shall be sufficient to meet the spirit of discontent, they disband their troops as soon as the immediate necessity for them has disappeared, often without giving them their stipulated pay, and thus themselves sow the seeds of new rebellion and mutiny before they have well got clear of the old. As long, too, as they are threatened on every side with insurrection, as long as they feel themselves weak, they will fawn upon, and grant privileges to, foreigners, which they will withdraw and turn into insolence and pride the moment they become free and untrammelled.

During the year 1866 there was scarcely any part of the empire of China which was not in some way the theatre of insurgent movement; and it has required very vigorous measures on the part of the departmental mandarins to suppress the outbreaks, for the Emperor deputes the task of putting down an insurrection in a distant province to the governor of a neighbouring province, and thus one governor is set to keep another in order. But the chief reason why these numerous insurrections do not succeed in their object appears to be the want of some leader who is capable at once of moving the multitudes to mutiny, and conducting them to victory—a want which probably arises out of the nature of things as above described. Otherwise the discontent, everywhere apparent among the people, would overcome the weakness of their rulers, and thus oppression would be punished by disaffection, rebellion, and a just and severe retribution whenever it became successful. The Emperor is a mere puppet in the hands of his nominal servants, and retains his position on the throne only by virtue of their forbearance, which again arises from their own selfishness and hope of aggrandisement; and the position of the Empire at the present moment, if we may be allowed to judge by the history of other great and ancient Oriental nations, is that of one tottering to its fall, or at least to its dismemberment. Human endurance can not be pressed beyond a certain point; and the corruption, oppression, maladministration, and tyranny of the Chinese government have reached a climax which cannot long delay its doom.

There can be no doubt that apathetic as the Chinese people are, and little as they appear to appreciate the civilisation of the West, intercourse with Europeans is exercising an important and silent influence upon them, which will one day make itself felt. The entire absence of intercourse with foreigners, which has characterised the Chinese nation for such a long series of years, has naturally imbued their minds with a degree of self-esteem and vanity which cannot be eradicated in a year, or even in a generation; but it is not in the nature of things that this should endure for ever. Their apathy has but one source, and that is ignorance. The foreign ministers of China, or rather its rulers in general, have hitherto had no interest in making themselves acquainted with foreign institutions or the status of foreign countries, and they therefore naturally remained in profound ignorance of these things; and, in the absence of information, it is no wonder that they believed themselves to be in all respects superior. With few exceptions, they have the most ridiculously erroneous ideas concerning us; and their notions of their own superiority are not put on to make an impression upon barbarians, but are the bonâ fide articles of their own candid belief.

But this state of things is insensibly changing, and since the treaty of 1858 it is not improbable that the Chinese have been the greater gainers of the two. For they have everything to learn from us; and although, of course, they may learn some evil, they must besides imbibe a vast amount of sound knowledge, useful and good. But we, on the other hand, only derive commercial advantages from the Chinese, in which they also largely share. No one can deny that

Lord Elgin's treaty was drawn up in a way which reflects the profoundest credit upon his qualities as a statesman and a philosopher. No treaty could have been more difficult to frame than that one which was forced upon an obstinate and half-civilised nation like the Chinese by a power which they regarded in the light of a hostile race of inferior, but temporarily victorious, barbarians, whom it was necessary to treat with cunning forbearance, but against whom they felt the most inveterate hatred. But Lord Elgin succeeded to admiration; and his treaty, drawn up with such care and skill that scarcely any ambiguity has ever been detected in it to cause a difficulty, has at the same time worked smoothly and harmoniously till the present time. But while the advantages to this country have been great in the interchange of commerce and the employment of capital, the results to those employed in direct commercial transactions have been by no means so satisfactory. Before the treaty of 1858 the trade of China was in the hands of a few great and wealthy houses, which, holding a monopoly of Chinese trade, regulated the markets and made immense profits upon the merchandise which they found at the ports where they were allowed to trade, and beyond which they knew little and cared less about the country. Their vast transactions, and the golden returns which they exhibited, invited numerous eager aspirants to break the monopoly, and to share in the commerce which made the few merchants truly princes as regarded wealth. The treaty did amply what these required, and China was thrown open to all, without restriction; but the results have not verified the dreams of prosperity which floated before the imaginations of those who were anxious to share in the benefits of free-trade. Competition, doubtless good in the main, has entailed much evil, the division of profits having failed greatly to benefit them; while the old stream of commerce, which enriched the few, has been nearly dried up, and decay and ruin have in too many instances been the only harvest that has been reaped.

But that Western influence is slowly but surely making itself felt is fully proved by the action taken by the Pekin College within a year of the present date. It was represented in a memorial to the Emperor, from the department of Foreign Affairs, that it was desirable that officials should be invited to pass an examination in astronomy and mathematics "with a view to the acquisition of a thorough understanding of foreign appliances." The memorial went on to state that whenever the usual path for the admission of candidates to the public service had been widened, talent had been called forth, and scholars of ability had eagerly presented themselves; and it called attention to the fact that in 1862 a school of languages had been established in the same department, in which English, French, and Russian teachers were assigned to every class; and explains that since the appliances of foreigners—their machinery and firearms, their vessels and carriages—are one and all derived from a knowledge of astronomy and mathematics, it is desirable that they should be learned, not superficially, but from the very foundation. The memorial then specifies the class of literate graduates who shall be competent to offer themselves as candidates for this new study, who are not to be under 20 years of age; and states that foreign instructors were to be engaged under the direction of Mr. Hart, Inspector-General of Chinese Customs, in virtue of his high official position; and they characteristically add, "The Chinese are not inferior in ingenuity, or cleverness and intelligence, to the men of the West, and if students (in the sciences of astronomy, mathematics, natural history, manufactures, mechanical appliances, *and the prediction of the future*) will so earnestly apply themselves as to become possessed of all secrets, China will then be strong in her own strength."

This is indeed a new era in Chinese history, and this is one of the most interesting and important documents which has ever been spontaneously issued by Chinese officials, since foreign intercourse has become an object of political importance. No pressure has been applied from without, but the influence of the Western representatives at the court of Pekin has in a great measure brought about this consummation, which its promoters defend argumentatively; and the fact that the memorialists risk unpopularity by taking such a step proves that they are convinced of the importance of Western civilisation, and are not entirely dead to progress and regeneration. There is but one paragraph in the memorial which can be otherwise than pleasing and satisfactory, and that perhaps is but of slight consequence—it states that "the germ of the Western sciences is originally borrowed from the Heaven-sent elements of Chinese knowledge. The eyes of Western philosophers having been turned towards the East, and the genius of these men being minutely painstaking and apt for diligent thought, they have succeeded in *pursuing study to new results*. For in reality the methods of their philosophy are Chinese methods—China has originated the method, which Europeans have received as an inheritance." This vainglorious boast will correct itself as Western science becomes instilled into the minds of the more intelligent Chinese; but the main fact to be remembered is that henceforth no man of learning in China will be able to proclaim his contempt, and boast of his ignorance, of Western learning—for that very learning will become his best passport to office, and a distinctive qualification which will raise him above his fellows.

This memorial has received the imperial sanction, and the thin end of the wedge has been fairly driven into Chinese prejudice and exclusiveness. And when it is remembered that it was only nine years after Lord Elgin's forced treaty, and only seven since a British conquering army entered Pekin, it speaks well for China, which up to that time believed itself the only country in the world worthy of imitation; and that everything appertaining to it —its language, laws, literature—were the sole fountains from which all the other benighted nations could derive benefit or instruction.

Those who have heard that it has lately been a moot point whether railways should or should not be forthwith introduced into China, will have a strong feeling of the rapidity with which innovation is gaining ground. But although for the present the Chinese Government has decided against the introduction of a railway system, it is no less remarkable that such a novelty should have been ever canvassed in high quarters, and the subject really argued, and not thrown aside as unworthy of consideration. It was proposed to make an experimental line, but Prince Kung met the idea by a sophism, in which he admitted the admirable character of the invention, and the benefits conferred by railways, and therefore argued that there was no necessity for constructing an experimental line to prove such well-known facts. But the real truth is, that all the objections which fifty years ago were urged against railways in this country, are now brought forward afresh by the fears of the Chinese. The abolition of old and time-honoured methods of travel—the destruction of the means of living of a large section of the population,—are urged against them. At the same time there are serious difficulties arising from the peculiar spirit and laws of the nation. Ground would of course be required, and so much is taken up with ancestral tombs, that to avoid them would be difficult, while to touch them would be to do violence to the strongest feelings of the people. Again, the Chinese could neither construct nor manage a railway by themselves, and therefore foreign aid and an English company must be called in to effect it for them. But in that case the occupation of a horde of peculating officials would be gone, and the whole constitution of society changed; the illegal perquisites of numberless small tyrants would be stopped, and they would be forced to starve upon their legal incomes. These are serious difficulties doubtless, especially when it is remembered that those who will thus suffer are those who are called upon to decide in favour of the introduction of railways. Another important point is, the universal system of barrier-taxation, which would be annihilated. Vehicles of all kinds, on river or on land, are squeezed at these barriers; but how can a goods train be stopped and examined in the same way? These prejudices have such show of reason in them, that it is no wonder that at the present time, and on the first proposal to introduce railways, very considerable opposition has been brought to bear, and with temporary success.

But the great barrier to a better understanding between the Chinese and the English is language,—not only the impracticability of Chinese to an Englishman, but the difficulty of English to the Chinese. Of course the comparatively few Chinese who are brought up as children in our schools learn our language with facility; but, although a barbarous corruption of the Queen's English is current in Hong Kong, it only serves as a very imperfect medium of business-communication,—a lame substitute for the genuine grammatical tongue. Nor is it more easy to teach a Chinaman good English, than for an English man of business to learn Chinese; the former can, on no account, be induced to learn beyond a certain point, just sufficient for mutual comprehension on very limited subjects, and there they inevitably stop. So also the Englishman, as a rule, would laugh to scorn the idea of understanding Chinese. Not that it is an impossibility, as some suppose; for while we all know the proficiency which a Morrison or a Legge has attained in the written character, so also I have heard an American gentleman address a native audience in Canton with *perfect fluency* after a few years' residence. But the difficulties are doubtless enormous. Few or none have ever mastered both the written and spoken language; but although the written character is necessary for the student, the colloquial (and more easy portion) is the available means of intercommunication. It is impossible for a Chinese to afford information on points of political economy, government, or literature, in *pidgin* English; and as it is equally impossible for more than one or two English to converse in Chinese, the two peoples are like two deaf persons conversing with one another, who may make a few mutual inquiries, but can never become acquainted with one another. The Chinese have given us a lesson in this respect, by establishing the school of languages at Pekin, with English, French, and Russian teachers; and it is really time that we should do our part and pay some attention to Chinese. The ignorance of the language is a frequent source of litigation and inconvenience in Hong Kong, where English merchants, perforce, engage Chinese assistants, whom they *secure*, the agreement being drawn up in English and Chinese, and the Chinese edition being often different from the English, neither party fully understanding the foreign version. A Chinese college is required to effect this great reform—this step in the regeneration of the great Eastern nation by the West; and such a college has been publicly proposed in San Francisco, where Chinese abound, and which carries on frequent and direct communication with Chinese ports.

As for the Christianising of China, that process must progress with very slow steps under the present *régime*. Ignorance of the language implies a most imperfect knowledge of their theories of religion, and unless we know them, how can we combat them, or ask them to substitute for them our own, which they can only most imperfectly comprehend? An isolated case of conversion may occur now and then, but the whole spirit of the two nations is so different, that nothing but a free interchange of thought can possibly tend to amalgamate them, by affording an insight into the great points on which they differ, or those on which perhaps they unknowingly agree. And without some such amalgamation of feeling, no great advance can be made in redeeming the Chinese race from the thraldom of a senseless paganism, such as must inevitably keep the national mind degraded and contracted, and the national morality at the lowest ebb.

CHAPTER XXI.

THE SURFACE POPULATION OF THE OCEAN.

Floating Animals—Capriciousness of their Appearance—Calms—The Towing Net—Medusæ—Nocturnal Animals—Formosa Channel—Hydrozoa—Yellow Fly—Blue Animals in Deep Sea—Abundance of Animals in Bad Weather—Lucernarian Jelly-fishes—Their Vast Numbers—Peculiarities—Portuguese Man-of-War—Stinging Powers—Fish Sheltering in their Threads—Sargasso Sea—Its Inhabitants—Atlantic Calms—Compound Salpæ—Three Forms—Chains of Salpæ.

DURING a long voyage, when the attention has been daily directed to the animate objects floating upon the surface of the sea, it must necessarily happen that much will be observed, and many interesting animals be met with. And when, moreover, that daily observation has been carried on for more than a year, and in seas of different latitudes, one can hardly fail to have noticed the greater number of the Pelagic creatures which habitually inhabit the surface of the ocean, as well as most of the phases of their appearance.

The numbers and variety of such floating creatures are very great, and by no means confined to one class of marine animals. There are certain fishes which habitually reside in the upper stratum of water, and are constantly taken by skimming the surface. Among these are glass-eels (Leptocephalus), Malthe, &c. Cephalopods (cuttles) of some species, as Spirula, not unfrequently occur also; and mollusks are common. These for the most part belong to that division termed Pteropods, from the wing-like aspect of their locomotive apparatus—and are either possessed of a shell or not, but when so possessed it is usually of most delicate structure and beautiful form. But there are other oceanic shells (as Carinaria and Janthina), of which mention may be made; and, moreover, some shell-less Nudibranchiata (as Glaucus and Scyllæa) occur in the same situation, as well as Tunicates, of which the luminous Pyrosoma is a good example. Floating Crustacea are not uncommon—either crabs of considerable size, as Neptunus and Lupea, or the numberless forms of Stomapods, Amphipods, Isopods, &c., which, with minute Entomostraca, constitute, perhaps, the bulk of the surface population of the ocean. Certain worms, also, are occasionally met with, and of Hydrozoa, the pelagic species are numerous and interesting, and will receive their share of notice. On one occasion, on the coast of China, a small Anemone attached to a piece of straw which floated it, came in among the produce of the net—a curious instance of the migratory power of a fixed animal; and it was no uncommon circumstance to find fixed Polyzoa and Foraminifera (Orbitolites) attached to floating leaves of Zostera or Sea-wrack.

There being such a vast number of animals whose normal dwelling-place appears to be at, or near, the surface, it would be readily imagined that scarcely at any time—at all events in calm weather—would the sea appear to be altogether without inhabitants. And probably there is, strictly speaking, no time when some living animals might not be found, if proper means are used to detect them; but it is no less true, that there are times when nothing is visible to the eye, although apparently all the conditions for their appearance are fulfilled.

Indeed, the influences which cause marine animals of the kinds enumerated above to rise to the surface and float upon the sea, would seem to be very obscure and capricious. For although it is a rare circumstance that the towing-net fails in securing some animals which would otherwise have escaped observation, it is nevertheless comparatively seldom that they are in such numbers, or so conspicuous, as to attract attention from the mere fact of their floating; and when they do so, it is not unfrequently under conditions which would at first sight strike the observer as anything but felicitous. Thus, when it is considered how delicate is the texture, and how fragile the structure of the majority of floating animals, it would at once appear that fine weather and a calm surface would be a combination of conditions most favourable to them. The ripples and waves of a disturbed sea would be, to all appearance, sufficient to mutilate or even to destroy such tender animals. And yet calms are by no means the only occasions on which they come to the surface; indeed, they exhibit a singular caprice in this respect. Thus, on one occasion, in lat. 12° N. and long. 58° E., when the sea was without a ripple, Porpitæ, and various other Acalephs (jellyfishes) floated in considerable numbers, with occasional Carinariæ, and the water was moreover alive with myriads of small Crustacea, which congregated in dense patches of a reddish colour. The sunshine lighted them up—like thousands of little sparks—as they rapidly darted about just below the surface. But, on the other hand, when in lat. 5° N. and long. 85° E., we experienced one of the most perfect calms it has ever been my good fortune to witness, in which every possible condition favourable to floating and delicate animals seemed to be fulfilled; yet not a visible speck broke the mirror-like smoothness of the blue sea from morning till night, excepting only the shoals of flying-fish, which from time to time relieved the somewhat monotonous scene with the life of motion.

Floating animals, then, are of such a character as to be either visible or invisible from the point of observation. The invisible ones are so minute, or so transparent, that they can only be recognized when taken in the towing-net —a bag of muslin or bunting with wide mouth, and which being let over the side of the vessel, skims the surface of the sea. The contents, indeed, are not unfrequently so delicate, that even when thus captured, and placed in a vessel of sea-water, they can only be perceived when held in a favourable light, or followed by means of some speck of colour which distinguishes them as they move along. The small Gymnophthalmatous Medusidæ (naked-eyed Medusæ) are of this kind, and may be very abundant, but would otherwise pass unnoticed; and it is seldom that the net is put down without securing various beautiful forms of such transparent Medusæ, as well as Beröes, &c., and small gelatinous masses, usually more or less torn by contact with the net. Many minute bodies doubtless pass through even the finest meshes; and it is evident that the net can only be used to advantage when the weather is fine, the sea tolerably calm, and the ship not sailing too fast.

I have never found the net entirely empty, excepting on one or two occasions, in the Singapore Straits; for although the nature and variety of its contents varied much, it most usually contained a great deal that was interesting and curious. There was not unfrequently a well-marked difference in the character of its contents by day, as compared with its captures during the night. Thus, for example, after being down for an hour or two one evening, I drew it up with a solid mass of minute Zoëæ; but although anchored at the same spot, in the produce of the succeeding night there was not a single Zoëa, but in their place transparent Crustacea (Leucifers), Entomostraca, &c. The glass-crabs also (Phyllosoma) always made their appearance in the night net. These curious little Zoëæ, now known to be the young condition of some species of crab, had enormous eyes, and grotesque helmets spiked before and behind. On the occasion referred to they appeared to be all of the same species, and nowhere else were they in such profusion, although sporadically met with, especially on the coast of Borneo.

I have elsewhere stated that the east coast of Formosa yielded perhaps the greatest variety of minute and inconspicuous, but at the same time highly interesting and curious, animals to the towing-net. Off Kackaou a single haul has produced a crowd of Entomostraca (minute Crustacea with a jerking locomotion), little Medusæ and Annelids; the Pteropod, Creseis; the tube-worm, Cerapus; blue Porpitæ; minute Globigerinæ; and numbers of little fat crab-like Megalopas, now known to be an advanced stage of Zoëa in the development of Crustacea. Sometimes rarer and more remarkable animals occurred, as the shelled Pteropods, Spirialis, Cleodora, and Hyalæa; transparent Firolæ, arrow-shaped Sagittæ, inert glass-crabs (Phyllosoma), elegant hyaline Crustacea (Alima and Squillerichthus, &c.), active shrimps of various degrees of transparency and minuteness, the oceanic nudibranch Glaucus, the spider-like marine insect Halobatis, the anomalous Pterosoma, with every now and then little greyish-yellow swimming crabs (Lupea pelagica), either side of whose carapace was developed into a long spine; and several minute fish, among them young flying-fish and Hemiramphi.

Besides this assemblage of animals, Hydrozoa (jelly-fishes) often abounded, more particularly pelagic species of the orders Physophoridæ, of which the Portuguese man-of-war (Physalia) is an example, and Lucernaridæ or umbrella-form Acalephs, like those thrown up on our own shores; of the former, perhaps, Velella, Physalia, and Porpita occur more frequently than any others, and usually in company with one another—the first two especially seldom seen one without the other. The Physaliæ and Velellæ look like large bubbles as they drift by at a little distance, but their persistence attracts attention, and their rich colours cannot fail to strike the most unobservant, especially when of large size. Exposing considerable surface to the wind, they sail along with the faintest breeze, and in a gale are huddled together in fleets, and stranded in great numbers upon the nearest shore. I saw thousands of both at the mouth of Kelung harbour, Formosa, after bad weather. The Porpitæ are less common, but usually occur in considerable numbers when seen at all—looking like beautiful and sharply-cut gun-wads, with delicate radiating markings, and surrounded with a fringe of deep-blue tentacles. The number of these three forms of Hydrozoa must be enormous, and their range very remarkable. I have found them extending over 55° of latitude, and have no reason to believe this to be the limit.

In lat. 12° N., near Socotra, and again in 5° N., near Ceylon, I was not a little surprised to observe in great numbers a small fly of a yellowish-brown colour, hovering over the calm sea, flying in gyrations near the surface, and occasionally settling upon the water and flying off again. I watched this insect, as I presume it to have been, with much interest, and was greatly disappointed at my unsuccessful attempts to secure a specimen, owing to the rapidity—10-12 knots—at which we were proceeding; but I saw them so often, and watched them so long, that I could not be mistaken in the fact. This little yellow fly I subsequently saw in the North Atlantic, between 30° and 35° N. latitude. While under steam in the calms, being occupied in the attempt to fish up some of the floating Acalephs and Ascidians, I repeatedly observed it settle upon the water, then rise and take a short erratic flight over the surface—but in vain did I essay to capture a specimen.

A notable circumstance occurred in the Indian Ocean in lat. 25° S., just south of the Mauritius. For several days in succession the net produced Halobatis,* glass-crabs, Velellæ, and the beautiful oceanic shell Janthina, of a rich, deep violet colour. But what struck me as very remarkable was that with the sole exception perhaps of the dark Halobatis, everything which the net contained was either transparent and colourless, or tinged more or less deeply with the rich violet of Janthina, which indeed nearly approached the sapphire-blue of the deep sea. There were small violet

---

* The occurrence of this singular hemipterous insect at sea is at least very remarkable. There appear to be several species, of which I met with two, one on the coast of China and the other some 500 miles from land in the South Indian Ocean. That they are veritable marine insects I think cannot admit of a doubt, though how they exist in the open ocean is a mystery. They are of a deep bluish-black, with six legs, the two hindermost furnished with a delicate brush on the inner side of the tarsus. The abdomen is remarkably undeveloped. Although taken occasionally in the towing net, I did not find them common, and never observed any movement after capture, owing to their delicate soft bodies being injured by the passage of water and other things through the net.

shrimps, little violet crabs, Physaliæ with violet blue threads, beautiful crystalline Crustacea, almost transparent, but tinged with violet. Small as these objects were, they would have escaped observation except for the towing net; but had they been larger, their colour so assimilated with that of the sea, that they would have been equally invisible from the ship.

In the Indian Ocean from Anjer to Natal, in April and May, although constantly on the watch I never *saw* a single floating object. This certainly appears strange, but, as before observed, the combination of apparently favorable conditions by no means always results in great numbers of floating animals. The reverse of this was curiously illustrated on one occasion when lying in the spacious harbour of Kelung in North Formosa, and although I have detailed the circumstance already, I must allude to it again here. The weather on this occasion was wet and boisterous, but nevertheless myriads of Creseis swarmed in the harbour, filling every mesh of the towing net, and giving the water a rippling movement and twinkling aspect, from the millions of little pairs of fins in constant motion. As the rollers came in from the north-east, great quantities of curiously-carved gelatinous Stephanomiadæ floated by, and as the afternoon advanced, and the rain increased, so also did these singular organisms augment in numbers, spite of the adverse circumstances which accompanied them. It was one of the very few days on which the sea might be said to be alive with curious animals; notwithstanding that there existed at the same time a combination of circumstances under which one would least expect to see such a phenomenon.

It becomes a curious question, whither go all these pelagic animals, whose home is the wide ocean, when they are not observed upon the surface? Why are they not more frequently seen? and why are the occasions so rare in which they are observed in such profusion? There must be circumstances connected either with their physical constitution or their modes of obtaining food, with which we are entirely unacquainted, but which must materially influence their movements. Doubtless they are sunk below the surface a short distance when not seen, for we cannot suppose that, short-lived as they may often be, they are suddenly produced like a crop of mushrooms in damp weather. They must exist somewhere, and a common influence probably regulates their movements, which perhaps need be but slight to bring them into view, or to carry them once more out of sight.

I never obtained the Pyrosoma in the towing-net, nor did I ever see them floating upon the surface. Yet these oceanic animals doubtless abound, and if I am right in attributing certain luminous appearances to them, they must most commonly float at a distance of two or three fathoms below the surface, though I have on one or two occasions seen the luminous body whirled to the surface in the eddy of the ship's wake.

With regard to the Hydrozoa of the order Lucernaridæ (the covered-eyed Medusæ of Forbes), on the comparatively few occasions when they appeared upon the surface, they were usually in great abundance, and not in great variety. Thus in the upper part of the Red Sea, on the 10th of March, a species of Aurelia appeared in great numbers; and two days after, we passed through a shoal of Rhizostomas. Four days later, in the Gulf of Aden, we again encountered shoals of Aurelia, apparently identical with those of the Red Sea, the two shoals being separated by about 1400 miles. Again, in October, we passed, on the west coast of Borneo, off

Cape Santubon, through a number of magnificent Pulmo-grades. The upper part of the umbrella was pilose, or hairy, with long papillæ; the circumference was fringed with slender tentacles, and the pedicels gave rise to magnificent grape-like masses, the whole being of a delicate white colour, and fully 18 inches in diameter. In the following month, in the strait which separates the island of Singapore from the Malay peninsula, I observed a great number of the same beautiful Pelagian, and accompanying it some specimens of a small and elegant, brown, torquoise-studded species, similar to one I had already obtained in Victoria Harbour, Labuan, and in which it may be here mentioned I found a small crab within the umbrella, beneath which it appeared to reside.

To show, however, the vast numbers of these animals which swim freely in the ocean, I will mention that, in the Atlantic, in lat. $3\frac{1}{2}°$ S. and long. $17°$ W., we encountered a shoal of Acalephs, all of the same species, the individuals of which were among the most beautiful in form and colouring that I have ever met with. They were of a delicate amethystine tint, speckled all over with a deeper colour; the umbrella was semitransparent, and the whole form wonderfully graceful. Just before sunset we passed through them for a space of two hours, during which time we had traversed ten miles. Supposing that this shoal were at least as broad as long, it was easy to calculate roughly that there could not be less than thirty millions of individuals constituting it, an estimate probably far below the truth. Well might Spenser exclaim in the "Faëry Queen,"—

"So fertile be the floods in generation!"

It occasionally happened that the observation of a shoal

of Hydrozoa pointed out some curious facts from which interesting deductions might be made. Thus, while passing through the Indian Ocean, in lat. 13° N., during an entire day (March 17), we passed through shoals of Aurelia, meeting from time to time patches in which they were too numerous to be counted, and in each of which there were many hundreds. A noticeable fact I remarked with regard to them, was that, without any exception that I could discover, these Aureliæ were, during the whole day, swimming in the same direction, or *with* the wind. We were steaming nearly due east, and a breeze was blowing a little south of east, and the umbrellas were all inclined one way, and pointing in the direction towards which the wind was blowing.

On another occasion, in a dead calm, on a beautiful day, off the river Min, I observed great numbers of a large white species. The edges of the umbrella were frilled, and numerous long and delicate threads stretched out straight and parallel; but what struck me as singular was, that these threads did not all float in the same direction, as though drifted from the animal by wind or tide; but although they were several feet long, they formed three or four distinct bundles, which stretched out straight, but in different and often opposite directions from the body of the animal, from which it appeared that they were propelled by a voluntary effort.

In passing through Banka Strait, owing to the number of rivers (the Palembang and others) which flow out of the island of Sumatra, the water was found to possess only seven-tenths of the saltness of the ocean; but notwithstanding this comparative freshness, I observed a number of large white Rhizostomas floating just below the surface, and apparently unaffected by this peculiar condition.

The only exception I met with to the rule I have mentioned (namely, that when Hydrozoa floated, they appeared in considerable shoals of one species only) occurred in the great calms which we encountered in the North Atlantic Ocean, in the first fortnight of July, and which extended more or less over upwards of a thousand miles, during which, on two or three occasions, I saw several species of Hydrozoa mingled with vast numbers of compound Ascidians. Some of them were new and strange forms of Beröoids, with lateral expansive lobes upon which iridescent bands of cilia were placed, and approaching in appearance the genus Eucharis of Escholtz. One of these is figured at the head of Chapter XXIV. Some were abundant, others but few in number, only appearing occasionally, and therefore very difficult to capture from a moving ship. One of these I did succeed in taking; but there were at least three or four species besides the Velellæ and Physaliæ.

The most magnificent specimens of the last-named richly coloured animals (Physaliæ) occurred in the Atlantic Ocean, near the Equator. On the 19th of June, in lat. 13° S. and long. 22° W., wind S.S.E., therm. 77°, bar. 30°·10, the sea was moderately calm, and from time to time during the day splendid individuals of Physalia pelagica sailed by, attracting attention, even when far off, by their large size and brilliant colours. They had the appearance of beautiful prismatic shells standing upright upon rich blue cushions, the shell being radiated from the base or cushion to the circumference, which was fringed with a rich and bright rose-colour. They were not in great abundance, but one would float by every five minutes or so.

The largest Physalia which I examined measured as follows:—

Extreme length of bladder . . . . 8 inches.
Greatest vertical circumference . . . . . 10¼ ,,
Height of bladder above water. . . . . . 2¾ ,,

But this was considerably reduced from the natural height; for the rose-coloured crest had collapsed, which would have added at least ¾ inch to it, making a total of 3½ inches in height above the water. I had judged them to be about 8 inches long, before I captured one, by the expedient of throwing into the water a piece of wood of ascertained length, which I carefully compared with the animal as it floated near it. No one on board the ship had ever seen such magnificent Physaliæ, although they had been at sea many years. Some thought at first that they had seen them as large in the West Indies, but they were fain to confess at last that the large one I measured exceeded the largest they had ever seen. I saw these large Physaliæ subsequently on more than one occasion, the last being in lat. 26° N., though higher than this somewhat smaller specimens occurred.

The stinging propensities of these Hydrozoa were not generally known, but were destined to make themselves evident at the expense of one unfortunate man. A boat happened to be lowered early in the day, and one of the crew, seeing a large Physalia float within reach, took it up with his naked hand. The threads clung to his hand and arm, penetrating to the axilla and down the side, causing the man to yell with agony. He was quickly brought on board, and as soon as he reached the deck, ran about like a frantic maniac, so that it took several men to catch him, and when secured and the proper remedies applied, he rolled about for a considerable time, groaning with pain. His arm was red, inflamed, and swollen, and remained so for some hours after the occurrence.

One circumstance in relation to these large Physaliæ struck me as being very remarkable. Each one as it floated by had beneath it what at first I took to be its mass of tentacles and polypites; but, on more close observation, I found that the appearance was due to a shoal of small fishes accompanying the hydrozoon under protection of its appendages. The fishes were of various sizes, from 2 to 6 inches long, transversely banded, and looking in the water precisely like the pilot-fish (Naucrates ductor). There were perhaps a dozen of these accompanying fishes clustered together beneath the bladder of each Physalia. Every Physalia had its cluster; but this peculiarity was observable, viz. that under small Physaliæ the fishes were small, while under large specimens they were correspondingly large, being in fact, always proportioned to the size of the man-of-war which they accompanied. Unfortunately I did not discover this curious fact till late in the day; and when the boat was down in the morning I was unaware of it, or I should have made a point of attempting to secure a specimen of so interesting a fish.

What the relation is which exists between the fish and the hydrozoon I cannot say; but this correspondence between the sizes of the two animals seems to indicate that the fishes do not capriciously select their protecting Physalia. It is known that certain fishes harbour in the threads of the larger Lucernaridæ, or umbrella-form Acalephs; but I believe they have not before been noticed to accompany the Portuguese man-of-war.

The presence of these fishes also accounted for a remarkable circumstance I had observed earlier in the day. A large

Albicore swimming near made a sudden dash apparently at a Physalia, but did not take it; returning, however, presently to the charge, it made a clean sweep, no trace of the Physalia being left. Doubtless it was the small fishes which accompanied it, rather than the Physalia itself, which stimulated the Albicore's attack.

In lat. 24° N. and long. $36\frac{1}{2}°$ W., we encountered the Sargasso sea, and with it that crowd of animals which feed upon those floating pastures. The Sargasso weed made its appearance in large patches, usually upon the surface, but sometimes, apparently sunk to some distance below it. It varied considerably in appearance—was sometimes dark-coloured, dense, and compact, and covered with berries—at others pale and lanky, and with few berries. The masses were round and shapely, and usually scattered somewhat indiscriminately over the surface, but occasionally a long streak of collected bunches extended as far as the eye could reach, in the direction of the wind. By hooking up masses of this weed many curious animals were obtained, of which perhaps the most abundant were small crabs (Planes linæana), many specimens being found upon each tuft of Sargasso; next to them in abundance was Scyllæa pelagica, an oceanic nudibranch of very peculiar form, of which usually there were several on each tuft of weed. Its general colour is a lightish brown, and its long narrow foot is well adapted for crawling along the stems of the Sargasso; and from its back rise two pairs of broad, somewhat rectangular processes, while a pair of large rounded ones spring from the head. When these animals were placed in a glass of sea water they immediately turned over by the weight of these processes, and sank to the bottom, having a most grotesque appearance—the two pairs of body processes looking like two pairs of hairy legs, and those of the head like a pair of long drooping ears, and the whole animal singularly like a long-haired Scotch terrier. The spawn of Scyllæa I often found twined in a narrow cylindrical coil in a very intricate manner around the leaves of the Sargasso, and enveloped in a gelatinous substance. Here and there a curious little fish (Chironectes), of various shades of a rich brown colour, lived on the weed—the pectoral fins bent in a singular manner, and looking precisely like hands, by which it grasped firmly on to the Sargasso. It was most curious to watch one of these fishes in a globe of water, where it lived a considerable time, and readily came to be fed with little scraps of meat. But here again a remarkable circumstance was to be noted—viz., that all the inhabitants of the gulf-weed were more or less of the same colour as the weed itself. The little crabs here were all brown of various shades—the fishes were brown—the Scyllæas were all brown—so that a uniform tint pervaded the whole, and it was difficult to perceive the animals in the weed, unless they were in motion. The Polyzoa and minute Zoophytes (Campanularia), as well as minute Annelids, which grew upon the weed, also afforded interesting occupation for the microscope.

I had been told of a large crab having been seen to swim past the ship; and I one day captured such an one in the towing-net, a species of Neptunus (N. pelagicus), which must be a terrific scourge to these populous communities. With considerable swimming powers, and pincers of a peculiarly trenchant character, such as usually accompany the flattened, paddle-form of posterior legs, this monster appears to wander from patch to patch of Sargasso, depopulating one of these floating pastures, and then making for another, which it clears in like manner—a very tiger of the seas. It was of a clouded reddish-brown colour.

On the 4th July, in lat. $30\frac{1}{2}°$ N., and long. 36° W., we first encountered an immense shoal of compound Salpæ (Salpa pinnata), which were no less remarkable from their interesting and most singular forms and structures, than from their abundance and the vast area over which they spread. In the water they were perfectly transparent, but for two pink linear bodies, and a yellowish brown canal, which seemed to gain brilliancy of tint from being seen through some depth of blue sea. These Salpæ were united in sets of various numbers by a rectangular gelatinous pedicle, which sprung from the inner side of the body, and met a similar pedicle in another individual. There was sometimes a single Salpa floating, sometimes two, three, four, and so on up to 11 or 12, which were united together in such a manner as to present the appearance of the carpels of an orange; and not unfrequently a second series of individuals was added outside the circumference of the original spheroidal mass. This outer series was never so numerous as the inner one; but consisted of from four to six individuals, united to the common centre by pedicles of a correspondingly increased length. They floated by in immense numbers, usually in an oblique position, but without any great apparent locomotive powers—certainly not so great as those possessed by umbrella-form Acalephs.

Upon withdrawing one of these clusters from the water, I found that the individuals were united by a knife-like edge at the extremity of each pedicle, which readily separated, so that the compound animal very easily became detached and independent. This accounted for the very various numbers in the groups which I had observed. Each individual was in every respect precisely like the other, and each consisted of a tough bag of transparent jelly, open at each end, through which the water freely flowed. The apertures were large and gaping, and opened alternately, admitting and expelling the water—the opening at the upper end of the animal, which *admitted* the water, being of a distinctly valvular character. In nearly every one of these animals I found a small crustacean (Hyperia), which swam freely about in the cavity of their body, and seemed perfectly at home there—not probably taken as a prey, but a voluntary tenant, which could swim in and out of the Salpa at pleasure. Few Salpæ were without one of these, which was distinctly visible through the transparent walls of the body—so transparent that when placed in a white dish the whole animal became invisible, but for the three coloured structures which they all contained.

Besides the single Salpæ I have mentioned, and which were evidently fragments of a compound animal, there was also a kind differing considerably from them, yet having a family likeness, and which Vogt considers the simple form of Salpa pinnata. It was very rare compared with the compound form, of a fish-like aspect, transparent, open at both ends, with apertures similar in form to the other, but with no process for joining on to other individuals. On each side of the body was a row of beautiful crimson linear bodies, occupying rather more than the middle third of the margin; and its anatomical details, readily observed through its glass-like walls, were very beautiful. I only managed to secure one of them, though the others were floating by in myriads.

Associated with the compound Salpæ were others of a smaller size, but in every other respect closely resembling them. These were of the size of a very small orange, and differed from the large forms above described chiefly in the fact that they were invariably perfect in their association, and of a perfectly spheroidal form; never floating in twos and threes, but always associated in numbers of eight or nine to twelve or thirteen—also never having an outer series, as in the larger kind; nor did I observe in any of them the little crustacean so common in the first-mentioned. But in form the smaller were mere miniatures of the larger; and in their anatomy I could detect no difference. I imagined that they were a young condition of those first described; but this supposition was considerably shaken by the fact, that of the myriads I noticed I never could see an intermediate form between the two. Their union by long pedicles, meeting also by a knife-edge in the centre, was more distinctly visible than in the first form, and the effect produced was a stellate arrangement of perfect symmetry and great beauty. This form is also considered by Vogt to be of the same species as the larger one, and thus we have the curious phenomenon of these compound animals—whose union of several individuals to constitute a compound mass is in itself one of the most remarkable circumstances met with by the naturalist—assuming two distinct forms in their united condition, as well as having a simple or celibate form, which never connects itself in compound masses, but always remains single, and retains its own proper individuality.

The number of these animals associated in one globose mass varied. In one I counted nine, in another ten, in another thirteen. The mouths of all did not, as far as I could make out, open simultaneously, and the upper mouth had precisely the same construction as that of the first-mentioned, viz., with a fish-like profile, the opening being produced, not by the movement of both lips, but by a moveable flap of the lower or inner enclosed lip, which alone acted. They moved slowly by a rotatory motion, but appeared also capable of projecting themselves along. All these gelatinous animals, although themselves transparent, when seen from above had a delicate green tint, and when deep down, this green tint became so intensified that they appeared absolutely luminous. But they were none of them luminous in the dark, as I satisfied myself.

For eleven days we passed through shoals of these compound Salpæ, during the greater part of which time it was perfectly calm weather. I did, however, see them beneath the waves, when a stiff breeze had raised a considerable amount of commotion in the sea. During these eleven days we had passed from lat. $30\frac{1}{2}°$ N. to lat. $41\frac{1}{2}°$, that is over eleven degrees of latitude, or nearly 800 miles, during the greater part of which they were thickly abundant. Nor were they alone, but were associated with numerous oceanic Hydrozoa of several species, some of which I have already alluded to as of new and strange forms, but which unfortunately could not always be captured from a moving vessel.

While watching these animals, I one day saw two magnificent objects, which I took to be clusters of chained Salpæ, and which were truly wonders of the deep. One of these consisted of five or six large Salpa-like bodies, forming an oblique line, each one of a bright and delicate green colour, and with a large rich ruby spot which shone in the water like carbuncles. The other was a long convoluted and delicate chain, which might be compared to a necklace of diamonds set with brilliant rubies, the whole waving gracefully in the currents of the water, just as though Venus had dropped her girdle as she rose from the sea. But however it deserved the name, it was not the rare Cestum Veneris, but undoubtedly a chain of Salpæ.* They were near one another, and seen nearly at the same time; and as it was impossible to get them, I immediately made a sketch of them from observation. Singularly enough, on a subsequent day, in a dead calm, I once more saw both these objects in close proximity, and was able to verify and correct my sketch. We were at the moment getting up steam; and although a boat was lowered for another object, by which an interesting discovery was indeed made, these beautiful creatures had drifted out of sight, and scarcely was the object of its being lowered attained when the boat was recalled, and we were going through the shoals of marine wonders, propelled by the screw at six or seven knots an hour. How I longed to be in a sailing ship, with no steam at command to hurry one on just as the most interesting moments arrived!

* For these two objects the reader is referred to the engraving on the opposite page.

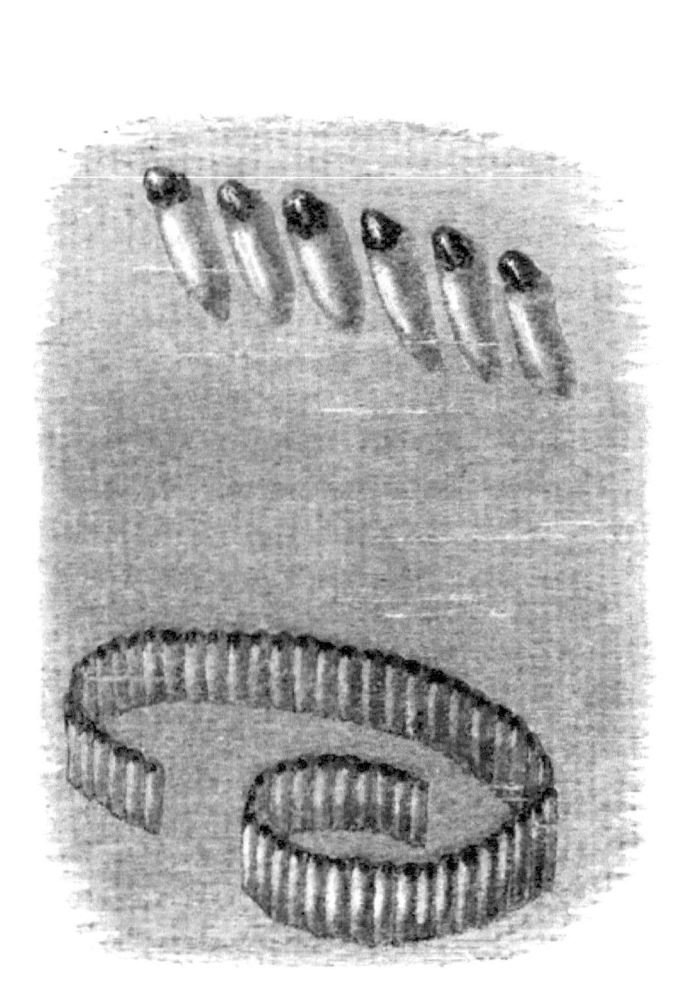

Chains of Salpæ

CHAPTER XXII.

OBSERVATIONS AT SEA.

Flying-fish—Their Range—Object of their Flight—Always away from the Ship — Mode of Flight — Absence of Vibration of Wings — Nature of Impulse—A Flying-fish Hunt —Albicores—Abundance of Flying-fish—Trichodesmium, or Sea Dust — Red Sea Conferva—Abundance of Conferva in the China Sea — Its Range — Cases of Red Discoloration — Microscopic Characters of Sea Dust—Oscillatoria—Observations of Former Voyagers — Horizontal Rainbow — Development and Peculiarities — Changing Aspect of the Sea— Natural Colour of the Deep Sea—Changes in Shallow Water—By Rough Weather—Father Secchi's Spectroscopic Observations.

ALTHOUGH few animals have been more often referred to than the flying-fish, and their habits described by many observers, the accounts concerning them are so conflicting that I was anxious to arrive at an unbiassed conclusion upon certain points respecting these interesting creatures, and I lost no opportunity of watching their movements with that end in view. I did not notice any flying-fishes on approaching the Equator, until reaching lat. $19\frac{1}{2}°$ N. in the Red Sea; but we afterwards found them as high as 26° N. in the Western Pacific in summer. In the Southern Hemisphere we lost them in $20\frac{1}{2}°$ S. in the Indian Ocean, and did not meet with them again till 14° S. in the Atlantic, (perhaps because it was winter in that part of the world), and finally parted with them in about 26° N.

The statements made regarding their mode of flight by

various observers differ so much that I wished to clear up, at all events in my own mind, any doubt or confusion which might exist on this point. The common impression appears to be that they emerge from the water either to escape from their enemies below, or out of mere wantonness; and that they disport themselves in the air for a certain time, which lasts as long as their wings remain moist, beyond which time they cannot maintain themselves above the water. But the difficulties of observation are quite sufficient to render it easy for a casual or inaccurate observer to be misled, and it is only after close and continued attention that I have convinced myself of one or two circumstances about which I was long uncertain.

In the first place, I became convinced that flying-fish never leave the water for their aërial journey without some real or imagined cause of alarm; that they never fly in the air to indulge their sportive humour, or to give vent to their exuberant spirits, but solely to escape from some peril which threatens them in the sea beneath. Their flight, therefore, is not that of a cheerful and happy animal indulging in a merry sport, as when unwieldy porpoises roll about on the surface on a summer's day; but it is the despairing and frantic attempt of a terrified creature to escape an imminent danger, which, though it may not always be successful, is more likely to be so in the case of such as are well provided with serviceable wings, than of such individuals as are not so well furnished. My reasons for this opinion are, that they always rise from the ship's cutwater or bow, and fly directly away from it; nor do they ever fly *towards* the ship unless palpably pursued by some voracious fish. A shoal will rise simultaneously from the ship's bows, and fly away in a series of straight but radiating lines, dropping irregularly into the water again, only to rise immediately once more, as simultaneously as if seized with a common panic; and thus they rise and fall two or three times, taking a course inclined about 45° to that of the ship, until ultimately a few only of the shoal may be seen emerging here and there at some distance on the ship's quarter. The only two instances in which I ever observed them fly against the ship's bows happened when they were pursued by Bonitos, or Albicores, a short distance off, when they became so terrified that they flew in all directions, blindly endeavouring to avoid their agile enemies; and I have been informed that under these circumstances they have sometimes flown in their terror so high above the water that they have fallen upon the deck of a ship. So also on two occasions I have known them fly into a cabin through the open port, attracted by a light burning within; but whether from the same kind of fascination which attracts the moth to the candle, or whether in consequence of the pursuit of Albicores, I am, of course, unable to say. Again, I may mention the fact that in situations in which we know flying-fish to abound, the day may pass without any being seen; nor, although the weather may have been fine and calm and inviting to the fish, did they emerge from the sea upon spontaneous excursions. Nor have I ever seen them flying at a distance from the ship, unless at a time when I could also see that they were pursued by Albicores, &c.

Next, with regard to their mode of flight. A shoal of a hundred or so will rise simultaneously,—some proceeding a considerable distance, say from one hundred to one hundred and fifty yards, without falling into the water, while individuals will drop after proceeding a few feet; and it is quite impossible in such a shoal to single out one for satisfactory observation. I found it best to watch for a single one, or two or three, and endeavour to follow their course; but the suddenness of their emergence from the water, and the rapidity of their flight, always *away* from the eye, made it difficult for a long time to detect the method of their locomotion. In general terms it may be said that they leave the water at a very acute angle, and, as a rule, not more than two or three feet above the surface, rising as the crests of the waves rise, and falling with their troughs, often touching the water, and, in many cases, dashing right through the tops of the waves without impeding their flight, lessening their speed, or materially altering their course. Moreover, no difference was visible in their speed and length of flight, whether they flew in the direction of the wind or immediately against it; so that the idea of their being borne along by the wind was out of the question. It was very evident, therefore, that the impulse of their flight was not all acquired before they left the water, for if so they would rise and fall at the same angle, and their course in the air would be in the form of an arc. Nor could they, under these circumstances, possibly fly a hundred yards without falling into the water. Indeed, whatever the primary impulse might be, it is evident that it could not carry them along over the surface of the water for any considerable distance, still less could it hurl them against the crests of the waves, and that in the very teeth of a strong breeze, without impeding their progress.

At the distance at which the flying-fish were from the eye before observation could fairly be brought to bear upon them, it was extremely difficult to detect anything like vibration of the wings, nor could it be said that it absolutely did not exist. But as some propelling power, while in the air, was absolutely necessary to account for all the phenomena of their flight, I was disposed to believe that a rapid vibration existed, similar in character to that of the wings of a fly; and I sometimes thought I could detect something of this kind in the change of prismatic colours which played upon their wings in the sun-light. But on one of the occasions above referred to, on which a flying-fish escaping from a Bonito flew towards the ship, I watched its approach, and saw it ultimately fall into the water immediately beneath me; and I was absolutely certain that the wings were in a state of perfect rest.

The opportunity of watching the evolutions of a somewhat larger species in the Atlantic, however, at last, I believe, supplied me with the clue which I sought. I then became convinced that every flying-fish, as it leaves the water, has its wings in a state of rapid vibration,—not so rapid as that the eye cannot follow them however,—and thus it gains an impulse in a horizontal direction. As soon as it is thus fairly launched, the wings assume a state of rest, somewhat in the position of those of a pigeon in the act of alighting, and thus they continue until the fish at length drops into the water. But when it meets, and is struck by, the crest of a wave, if it emerges from it immediately, as frequently happens, it does so with a similar vibration of the wings to that with which it first left the water; and each time it strikes a wave a new vibration succeeds, as though the contact of the water produced an automatic vibration of the wings which kept them above the surface. Each contact with the water, then, is followed by a vibration of the wings, producing a fresh impetus; and in their lengthened flights over smooth water, I early remarked that they occasionally *touched* the water in their progress, the touch being probably provocative of a new vibration which would carry them forward, although too far off to be observed. But when the fish meets with a succession of wave-crests, it takes a more or less zigzag course, changing its direction each time it emerges from the water, and at the same time waving from side to side after the manner of the sailing flight of a large sea-bird.

Crossing the Equator on June 19th in long. 22° W., I was witness to a remarkable scene, in which the poor flying-fishes played a conspicuous part. The whole day long the path of the ship was beset with a number of large fishes (Albicores), which played sad havoc among them. The Albicores were about 5 feet long, extremely active and bold, darting to and fro under the cutwater, and raising the flying-fishes in terrified shoals. Every now and then they would leap in a graceful curve 8 or 10 feet out of the water, and on several occasions one would make a succession of such leaps among the shoals of flying-fish, and, singling out one, catch it in the air, the victim being distinctly seen between the jaws of the monster as he fell into the water. It was an exciting and interesting scene to witness the leaping and splashing of the great Albicores, which pursued their prey with the rapidity of an arrow, and the frantic efforts of the flying-fishes to escape, which were often ineffectual. Leaving the water as usual in simultaneous shoals, they fled before their enemies usually at right angles to the ship's course, the wind being abaft; but I remarked, as an unusual circumstance, that after flying a little way they always veered round before the wind, so that their flight was almost universally bent in the form of a boomerang. Accompanying the Albicores were two smaller red fishes, which did not leave the water, but which were evidently in pursuit of *small* flying-fishes. I several times saw rise before them a little flying-fish, not more than 2 inches long, such as I had never seen before, and which only seemed capable of propelling itself a yard or two through the air.

At ten o'clock in the morning we hove-to, for the purpose of practising with shot and shell at a target; and for an hour the boom of 68-pounders, and the crash of 110-pounder Armstrongs, added to the splash of the shot and shell, might have been imagined to be sufficient to frighten the Albicores, &c., away. But although I did not notice them as long as we were lying-to, no sooner did we continue our course than the scene was resumed with renewed activity. One of the Albicores seized a hook baited with a rag, which was hung over the bowsprit, but the line was not strong enough for his violent plunges, and he carried it all away. They did not leave the ship all day; and in the afternoon I noticed two of them swimming close beside the bridge. They accompanied us at least two hours, always in the same spot, keeping pace with the ship—every now and then diverging after a passing flying-fish, but returning again to their station beside the bridge. How long they had been in that position before they were noticed I cannot say.

Flying-fish must be extremely abundant. On some days the shoals seen on the wing must have amounted to many thousands; and even when none were seen, proofs sometimes existed of their great plenty. Thus, when lying for several days on the edge of the Pratas Reef, in the China Sea, not a single flying-fish was observed on the wing, yet when I went among the gannets' nests upon the island, I found that every bird sitting upon the nest had four or five large fresh flying-fishes in its stomach, which it disgorged before taking wing. These were probably taken in the water by the birds. At Kelung, in North Formosa, I saw great quantities of flying-fish of a large size (about 18 inches long), which had been dried and packed in barrels, and were probably intended for exportation to Amoy.

I have hitherto purposely avoided speaking of a floating object which naturally attracted considerable attention on various occasions—not on account of its novelty, for it has been written about over and over again, and referred to by many travellers—but on account of its abundance and frequency in the seas in which my voyages were chiefly made. This substance is what is known to mariners under the various names of spawn, whales'-food, sea-sawdust, and other terms which are equally incorrect, for the substance is really a confervoid growth, to which the generic name of Trichodesmium has been applied by Ehrenberg. It has long been believed that the Red Sea derived its name, which signifies the same both in Latin, Greek, and also in modern Arabic, from an occasional red discoloration of its waters—a discoloration which was first observed by Ehrenberg in 1823, when spending some time on the coast of the upper part, and carefully examined by him, as well as by Montagne, by means of specimens supplied to him by an observant traveller. Both these *savans* agree that the substance discoloring the waters of the Red Sea is of a vegetable nature, being in fact a filamentous Alga, which has received the name of Trichodesmium erythræum—of a blood-red colour—which often covers large areas, and appears and disappears somewhat capriciously. Observers in other parts of the world have met with a copious deposit from time to time which appears to be of the same nature; and although it does not appear that this deposit is always of a red colour, it has been referred by competent botanists to the same species, which has since been renamed T. Ehrenbergii

—because a second blood-red species, T. Hindsii, had been found off the west coast of S. America. It is somewhat singular that both these species should have the characteristic red tint—a tint which I have never seen assumed by any of the vast quantities of Trichodesmium which have passed under my notice.

I saw none of this red appearance in my passage down the Red Sea, nor any indication whatever of the existence of Trichodesmium therein; but first observed it in the Indian Ocean in lat. 5° N., and long. 70° E., when I entered the remark in my journal that the sea had a *dusty* appearance, as though myriads of minute bodies were floating in it—an appearance rendered very distinct when the sun shone upon it. I fancied these motes might be minute animals, which perhaps produced the luminous sparks so often visible at night; but having succeeded in getting some of the water, I found that the objects in question were little bodies, which under a lens presented the appearance of sheaves of minute fibres, constricted in the middle, but loose at the ends, like sheaves of corn in miniature. It was not, however, till we were east of Singapore, and fairly in the China Sea, that this peculiar phenomenon became visible in all its remarkable features. Nearly every day, while traversing this sea, more or less of Trichodesmium was to be seen, and not unfrequently the sea was covered with a thick scum like that which settles upon a stagnant pond, only of a yellowish brown colour. In very calm weather this scum formed a regular, smooth pellicle in the water, thrown up here and there into folds and rugosities. Such a scum would sometimes cover the sea more or less for nearly the whole day with little interruption. If, however, a moderate breeze was blowing, and the sea raised, instead of a uniform pellicle, the dust would be arranged in long irregular parallel lines, or bands, extending unbroken as far as the eye could reach, and taking the direction of the wind. On one occasion we crossed a single band of this character, the only one seen on that day.

The frequency of this appearance in the China Sea may be judged of by the fact that out of four times that I crossed that sea, I observed the sea-dust to be more or less abundant during three of them, and assuming one or other of the appearances described. The fourth time was in winter (December), and during the height of the monsoon, the wind being very boisterous and the sea very rough, so that this substance was doubtless so thoroughly washed and dispersed by the waves as to be indistinguishable amid the turmoil and foam. The most northerly point at which I observed the accumulation of Trichodesmium forming a pellicle upon the surface, was at the north entrance of Formosa Channel, in lat. $25\frac{1}{2}°$; and it is somewhat remarkable that I should have seen none south of the Equator in the Indian Ocean, Rhio Strait being the most southern locality. On one occasion indeed, in lat. $28\frac{1}{2}°$ S., it manifested its presence by the same indications by which I first noticed it, namely a scintillating of the scattered sheaves below the surface—a fact which I proved by examining the water; and in the Atlantic, in lat. 80° S., the same appearance was closely followed by two or three bands or streaks, in which it was quite dense, discoloring the water.

But it is worthy of notice that on all these occasions the colour of the Trichodesmium was the same, viz. a yellowish brown, and never at any time red, or approaching it—much less the *rouge de sang* of the French botanists. On only two occasions did I ever observe the sea discoloured by red matter—once by myriads of minute red Crustacea, in the Indian Ocean, and again by a dense mass of red gelatinous worms in Formosa Channel—but never by Trichodesmium.

The characters presented under the microscope by the specimens I first obtained in the Indian Ocean have been already alluded to; and although the ultimate elements of the various specimens were the same, I met with two distinct forms of Trichodesmium, one on either side of the Malacca peninsula. That on the west side was in the form of a miniature sheaf of corn, while that of the China Sea was a cylindrical bundle of fibres, more or less pointed at one end, but obliquely truncated at the other. This was also the form it assumed on the other occasions on which I examined it. Both these, however, consisted of bundles of cellular fibres of the same character. The bundles were cream-coloured and opaque, and a lens showed that the ends were fimbriated, owing to the component fibres being loose at their extremities. With slight compression these fibres were seen to be cylindrical filaments of unequal length, combined together and interlacing one another, forming an intricate net-work, which resembled unfinished basket-work with the long ends of the osiers sticking straight out. Each filament was long, symmetrical, and unbranched, with a rounded extremity, and even, hair-like outline—divided by transverse septa into rectangular cells half as long as broad; and each cell contained some grains of chlorophyll in the centre, which rendered it opaque. Continued pressure, however, discharged part of this substance, rendering the cell-walls distinct, and ultimately the filament broke up into its component cells, which presented various facets to the eye—some round and some rectangular, proving its confervoid character—being in fact composed of a linear series of tubular cells.

No movement took place in any of these cells; but mingled occasionally with this was another microscopic vegetable, spherical in form, and bristling with minute rays like a miniature Echinus—and about as large as a pin's head. This proved under the microscope to be entirely different in character from the confervoid just described, and was indeed an Oscillatoria. It possessed also a gelatinous envelope, which I never could find in the Trichodesmium, but was in much smaller quantities than the latter.

The naturalists of Captain Cook's third voyage observed the substance I have described about New Guinea, and the sailors called it *sea sawdust*. Peron saw it extending for 20 leagues from east to west, of a greyish colour. Darwin noticed it near the Abrolhos islets, and says he met with an allied, but smaller, species off Cape Leeuwin, Australia; but none of the observers appear to have looked upon it as an every-day phenomenon, as it certainly appeared to be in the China Sea. Moreover, although it has been settled that the conferva observed by these travellers in different parts of the ocean is of the same species as that which discolours the Red Sea, I am myself very strongly of opinion that this is an error, and that it will be found that several species of this remarkable little Alga exist in different parts of the world. The two forms I have described are both of them in many respects different from Trichodesmium Ehrenbergii —nor have they much in common with the second recognised species, T. Hindsii.[*]

---

[*] For a more complete account of this substance, the reader is referred to a paper recently read at the Royal Microscopical Society, and which will be found in the "Microscopical Journal" for April 1868.

I have alluded in some parts of this volume to remarkable atmospheric effects occasionally met with in the open ocean; but none were more singular than the horizontal rainbow which I witnessed on May 5th, in lat. 25° 19′ S., and longitude 54° 13′ E. The weather was very fine and bright, and we were sailing gently along with a light breeze, when I observed signs of a squall blowing up from the S.S.E. I was sitting reading on deck at the time, and immediately went over to the port gangway to watch its approach. It was about half-past one p.m., and the sun was therefore in the N.N.W., exactly opposite the approaching squall, upon which was already developed a rainbow of low altitude (12° to 15°). While gazing at it my attention was arrested by a yellowish-brown haze upon the horizon immediately under the centre of the arc, which, although very faint, appeared from its position to have some connection with the squall or with the rainbow; and I was thus induced to watch it attentively. At first it was a mere indefinite tinge of colour on the distant horizon, and for two or three minutes it seemed to undergo no change; but at length by slow degrees it increased in intensity, and then appeared to spread over the water, looking as though a cloud of reddish dust was hanging over the sea. For some minutes I was quite at a loss how to account for it, but carefully watched to see what would be the upshot. It now rapidly intensified in brightness, and presently became prismatic; then slowly spreading forward across the sea towards us, it at length presented the appearance of a brilliant horizontal bow lying upon the sea, its apex just capping the horizon, and its limbs seeming to fade away upon the water halfway between the eye and the horizon. As the horizontal bow increased in intensity the vertical one gradually faded away, and quite vanished immediately after the former had reached its greatest brilliancy, which was most marked about the centre or apex.

In the horizontal bow the red colour was upon the outer or convex side, while in the vertical bow the red was on the inner or concave side. The *horizontal* bow was therefore the *primary* bow, and the vertical bow the secondary or reflected one. Hence we had the remarkable spectacle of a secondary bow appearing before the primary bow was at all developed, and fading in proportion as the latter reached its greatest intensity. The vertical bow, however, was always much less bright than the horizontal bow ultimately became. This latter, when once the prismatic colours became fully developed, seemed rapidly to approach us from the horizon, the limbs appearing to shoot forward, becoming broad, and spreading a wide coloured space upon the blue water on either side; and the bow, when complete, had somewhat of a horseshoe-shape, as though foreshortened. When it had reached its greatest intensity, being then of amazing brilliancy, it suddenly faded and disappeared, and the vertical bow, which had been growing very faint, disappeared at the same time. Throughout the whole duration of the phenomenon the apex of the horizontal bow maintained precisely its original position upon the horizon, namely, where I had first been struck by the appearance of the luminous haze; and from the time I first observed this appearance till the moment when the whole vanished was about ten minutes. During this time a small drizzling rain fell, which was scarcely sufficient to wet the deck, and the squally effect passed away to the south-west.

But while it might happen that no peculiar phenomena attracted attention in the sky, nor any living animals were visible in the water, it was always interesting to observe the ever-changing colour of the sea. The surface of the ocean is not the monotonous plain which some would make it out to be; it is ever varying with a succession of aspects both of form and colour. Now it is smooth and glassy, now breaking into dimpled smiles, the ἀνηριθμον γελασμα of the dramatist, now capped with foam and breaking all around into *white horses*, and now rolling in majestic billows, which I for one never tire of watching, as they bound along from afar off as though they meant to engulf the ship, and then raising her gently up to their highest crest, poise her above the boiling plain, and as gently lower her again into a smooth hollow valley, the emerald sides of which are streaked with foam. The sudden and rapid changes, and the ever varying prospect, form as near an approach to the wavy, skimming flight of a sea-bird as can well be imagined.

Nor is the colour of the sea more monotonous than its other aspects. Now a pale sapphire blue, it deepens into ultramarine, and then again into intense indigo, or blue-black. Again it may assume a pale yellow-green colour, and become bright emerald; and when the setting sun bathes the clouds with gold, the sea partakes of their glory, and dazzles the eye with a flood of light, which fades away like the dying hues of the dolphin through shades of purple and rose, until it once more assumes its twilight tint of deep indigo-blue.

The natural colour of the deep sea when perfectly at rest, in fine weather, is a rich violet blue, of an intensity and indescribable brilliancy which no pigment ever equalled. Nor is this colour in any degree dependent upon the blueness of the sky in the way of reflection, which not only does not cause, but in no way assists in intensifying the blue of the sea. For not only is this colour of the sea to be observed when the sky is cloudless, but also when, although bright, scarcely any blue sky is visible, the whole vault being filled with rolling *white* cumuli. But anything which intercepts the light of the sun changes at once this rich violet-blue into some other colour. If it is a passing cloud, or the shadow of the ship, while all around is bright, the sea becomes under its influence indigo-blue; whereas if direct light is altogether excluded, as on a dull cloudy day, the sea becomes of a deep blue-black, or even leaden hue. Thus I have seen it lead-colour and of a bright blue within a space of two hours, when the weather has changed from dull to fine. The same effect is also produced when the sun gets low, although it may be clear.

Near the shore, or in soundings, as it is expressed, the sea is never of this rich violet-blue, probably because the depth of water is not sufficient for the light to have its full and true effect. Moreover, the sea being beautifully clear and transparent, at a moderate depth the nature of the bottom has a perceptible effect upon the colour of the water. Usually under such circumstances the colour of the sea is olive-green, a colour which I have observed extending for 70 or 80 miles from land off the south coast of Africa; and nearer than this, when the water is shallow, it often has a variegated appearance, directly due to the various growths of weed and the irregularities of the sea-bed.

But even in deep-sea the water is not always violet-blue or indigo; but under the same conditions of light, the smoothness or roughness of the surface is accompanied by gradations of colour between blue and yellow. Thus on a fine day, such as, if also smooth, would have produced the characteristic violet-blue, the surface being ruffled, a fine, *light*-blue tint was everywhere visible, but more usually a shade of green; a circumstance which was particularly marked when, after squally weather, we were in a latitude in which the soundings marked in the chart were 2350 fathoms, but the sea was of a light-green colour; and the remark I have entered in my journal was that for some two or three days past, during windy weather, the sea has lost its blue colour, and to-day seems *washed out*. This peculiar phenomenon I attribute to the commotion which the sea has undergone, having entangled air with the water; and although no masses of foam are anywhere visible, myriads of minute air bubbles, mingled with the water, modify the usual absorption of light, and reflect more or less of the yellow rays. That this is the proper explanation is confirmed by a fact I have more than once noticed, viz. that when, in fine weather, the sea has been of the ordinary dark-blue colour, the wake of the ship has been marked by a path of light-green water for a long distance behind.

That blue is the natural colour of the water is moreover proved by the fact, that whatever the colour of the sea under the changing influences of light and shade, whether dull and leaden, or bluish-green, the water in the screw-well,—upon which we look directly down, and which is liable to no lateral reflection or disturbing influences such as the open sea must of necessity be subject to from the angle at which our eyes regard it,—is always *blue*, sometimes pale, sometimes dark; but under the most favourable circumstances, of an intensity which frequently attracts the admiration of those to whom it is an every-day occurrence. I think it was Sir Humphry Davy who attributed this blueness of the sea to the presence of *iodine;* but I cannot help thinking that it is an inherent property in the water, just as some shade of the same colour appears to be an inherent property in the air of the atmosphere, that is to say, that sea-water, as such, in sufficient quantities, absorbs the red and yellow with their compounds, rejecting the blue and violet for the benefit of our eyes.

Father Secchi of Rome, in examining the water of the sea, found that the spectrum of its colour lost its red rays first, and at a slight depth; after which, as the depth increased, the yellow and green rays successively disappeared to a great extent, and the water assumed its violet-blue tint. And in some researches which he has lately been making upon the colour of light transmitted through masses of ice, he finds that the red rays are very faint, being mostly absorbed, while the richest blue was transmitted, so that in a grotto of ice the flesh-tints of the human face assumed that ghastly appearance which is seen when a homogeneous blue light is burning; and he comes to the conclusion that the true colour of water is blue, mixed with violet.

CHAPTER XXIII.

THE LUMINOSITY OF THE SEA.

Nature of the Phenomenon—*Phosphorescence* a Misnomer—Classification of Luminous Phenomena—Sparks always visible—Their Cause—Luminous Sheath to Ship—Singapore Harbour—Simon's Bay—Noctilucæ—Scene on the Chinese Coast—Moon-shaped Patches of Light—Not caused by Medusæ—Often spontaneous—Probably Pyrosomas—Recurrent Flashes—Colour and Appearance spontaneous—Depth of the Animals—Examples of Recurrence—Milky Sea; its Rarity—Conditions of Luminosity—Non-luminous Animals—Rationale of Luminosity—A Correlative of some other Force—Contractility—Luminous Envelopes—Range of Luminosity among Animals.

THERE are few subjects of study more interesting than the luminous appearances presented by the sea under various circumstances, and the least observant person cannot fail to be struck with the remarkable phenomena which in the course of a long voyage he must perforce sometimes witness. That the sea, the great extinguisher of fire, should be turned into flame—that the darkness of night should be illuminated by the luminous glow which bathes every ripple, and breaks over every wave—that globes of light should traverse the ocean, or that lightning flashes should coruscate no less in the billows of the sea than in the clouds of the air—are all facts which seize upon the imagination, and enforce attention and consideration. Nor is the interest lessened by the knowledge that all these phenomena are produced by animals, whose home is in the great waters—that not only do the fiery bodies of large animals give out steady patches of light, but that of the myriad animalcules with which the sea teems, like motes in a sun-beam, each contributes its tiny scintillation, the aggregate forming a soft and lovely radiance.

The luminosity of the sea, its appearance, and its numerous forms—the various conditions under which it became manifest—and, as far as practicable, the causes which produced it, were subjects to which I was anxious to pay especial attention. For, although some of these points had already engaged the attention of competent observers, who have elicited much curious and valuable information, few of them have had the opportunity of watching the phenomena for a long-continued period, or over a wide extent of ocean. Much therefore undoubtedly yet remains to be learned regarding them; and I shall in this chapter collect together the various scattered observations which I carried on at every available opportunity during the space of a year and a half. Not a night passed while I was at sea without my looking out for luminous appearances, jotting down anything novel or unusual, and, whenever practicable, making an examination for the determination of the cause and *modus operandi* of the luminous manifestation. And although the bright moonlight nights were often very beautiful, I not unfrequently bewailed the invisibility of the luminous animals whose light was temporarily extinguished by the superior effulgence of the lunar rays; and I longed for a return of the dark, but no less beautiful star-light nights, the brilliancy of which compensated for the absence of the moon, without putting a stop to my observations upon the luminosity of the sea.

Before detailing the remarkable phenomena which presented themselves to my notice from time to time, let me say that I purposely avoid using the word *phosphorescence* when speaking of these appearances—a term very generally, and at the same time very loosely applied by most observers, but which has no right to any place in the descriptions of luminous phenomena as exhibited by the sea. Phosphorescence is here a misnomer, and an even greater misuse of terms is it to speak of phosphorescent *matter*. There is no phosphorus in the case, nor anything allied to it, except in the abstract meaning of the word; neither is there any combustion; but the light is, *sui generis*, the product of causes of an entirely different category from those which have to do with the light-producing properties of phosphorus. Nor is the light of a material character, such as could be spread upon the end of a match, like phosphorus; and although in some few cases the luminosity has appeared to cleave to extraneous substances, there can be little doubt that in such cases the light had a different origin, and was of a different character from the ordinary forms of animal luminosity exhibited in living organized bodies.

But we will first state the facts, and draw some conclusions afterwards. And it will make the subject clearer if we follow a methodical arrangement, and group the facts in an orderly manner. I would therefore classify all the cases of luminosity which have come under my observation, under the following five heads:—

1. Sparks, or points of light.
2. A soft liquid, general and wide-spread effulgence.
3. Moon-shaped patches of steady light.
4. Instantaneous recurrent flashes.
5. Milky sea.

The first of these, or the appearance of points or sparks of light, is by far the most common, and in different degrees may be said to be all but universal. Whether the other forms of luminosity are exhibited or not, sparks of light in greater or less abundance are scarcely ever absent. The sea, more particularly when agitated, sparkles with brilliant points of light, varying in size from that of a pin's head to that of a pea, and of greater or lesser permanency—some being almost instantly extinguished, while others retained their light for an appreciable length of time. I do not think I ever looked at the sea on a dark night without seeing some few sparks, even though I might have entered a remark that the sea was "*not* luminous to-night." But usually these sparks are abundant, and on occasions they present a wonderfully brilliant appearance. On one occasion, when this phenomenon was unusually striking, on the coast of China in lat. 26° N., on drawing up bottles full of water, and pouring it out in the dark, the water sparkled brightly as the luminous points ran over; but a close inspection revealed nothing in the water but a few minute Entomostraca. On another occasion, when some water which had been left in a basin exhibited luminosity at night, I got a very brilliant spark upon my finger, and taking it to the light, it proved to be a minute crustacean of the same division. I do not mean to say, however, that these sparks, when thus appearing as distinct and segregated scintillations, are always due to Entomostraca. There are many other minute creatures which exhibit luminosity; but I wish to draw a distinction between this form and that next to be described, which appears to be mainly due to one organism, which, owing to its occasional great abundance, produces phenomena conveniently distinguishable from this common and almost universal one.

The second form of luminosity then to be noticed occurs comparatively rarely. It consists of a soft, usually greenish light, which only makes its appearance when smooth water is disturbed, and is only seen in calm weather. This indeed appears identical with what we see nearer home, as on the shores of Ostend and in the estuary of the Mersey. This form of luminosity I have observed on only three occasions, but each time under similar circumstances; and I have reason to believe that the cause is the same on all occasions, whether in the eastern seas or in the Mersey. On the 5th of July, being on the coast of China, in lat. 27°, the weather in the afternoon became dead calm, and after sunset I remarked that the sea was beautifully luminous, but altogether without conspicuous sparks or points of light. Wherever the ripples caused by the advancing ship rolled away, they were crested with bright green light, and the ship's hull appeared to be enveloped in a luminous sheath. On this occasion the effect did not last long, and I did not examine the water microscopically.

The next time I noticed this form of luminosity was in Singapore Harbour, on November the 6th. The wind was east, thermometer 76°, weather fine. The water was like glass, smooth and beautiful, and exhibited no light except when disturbed; but every oar-stroke of the boat in which I was rowed produced eddying circles of brilliant light, and a lovely soft green glow crowned every ripple from the bows. The scene was perfectly fairy-like. As we pulled among the shipping, under a brilliant tropical star-lit sky, we left a fiery wake which widened behind us. Every splash in the water was like a shower of diamonds, and a myriad of minute sparks leaped up when I took water in my hands and poured it back into the sea, and the aggregate of these multitudinous and brilliant scintillations made up this delicate luminosity, which I never saw so beautifully exhibited as upon this night. The following night the same effect was visible, but scarcely so intense as before (wind N.E., temp. 76°), and on the third night (the wind being E., and temp. 75°) I again observed it. After this I was absent from Singapore two nights, and on my return I no longer noticed the luminous effect.

On each of these three nights, I carefully examined the water. As I filled a bottle, bright sparks of light adhered to my hands, or on bringing it to the lamp I found that it contained a number of small globular greenish bodies, which floated upon the surface for the most part, but appeared to have the power of freely moving in the water. On closer examination these bodies proved to be Noctilucæ; and during the night I observed that the contents of the bottle frequently flashed with bright and rapid coruscations. I had no difficulty therefore in coming to the conclusion that the peculiar luminosity in the harbour was due to the presence of innumerable Noctilucæ.

On the 24th of May, lying in Simon's Bay, Cape of Good Hope, the water was again luminous, in a manner similar to the occasion just alluded to. The weather was fine, wind W.N.W., light; bar. 30·04, therm. 60°. On examining the water closely, I found that, as before, the luminous effect, though soft, subdued, and apparently uniform, was really due to innumerable small sparks; and on bringing the water to the light, I found numerous Noctilucæ in it precisely similar to those observed at Singapore. They were not, however, in sufficient numbers to have produced all the light, for in a wine-glassful of water there were on an average not more than a dozen Noctilucæ. But besides these bodies there were a great number of motes in the water, many of which appeared, by their rapid jerking locomotion, to be minute Entomostracous Crustacea. They were so minute that, by the imperfect light on board ship, I long tried in vain to secure one to place under the microscope. Besides these were also some larger species of Entomostraca.

The Noctilucæ may be described from these specimens:— they measured from $\frac{1}{27}$, to $\frac{1}{335}$ of an inch in diameter; they were somewhat kidney-shaped, and of a pale greenish colour when seen with the naked eye, closely resembling Volvox in appearance, but with a much less active movement. They had, however, powers of locomotion, though the means were not apparent under the microscope. They had a dark nucleus, usually irregular, but in some cases spherical and well defined. The circumferential outline seemed very faint (on account of their globular form), and their general aspect was very variable. A kind of slit appeared to extend through two-thirds of the body, from which faint lines radiated, usually having a double outline, and not reaching the circumference of the sphere, but often terminating in large, round, granular bodies of various sizes. The whole body was studded with minute spherical interspaces (vacuoles) of various sizes, which strongly refracted the light, like oil globules; but slight movements, which appeared to be taking place in an almost imperceptible manner, soon changed the whole aspect of any individual Noctiluca while under observation, so that the description or drawing of one minute did not answer for the next. Each Noctiluca had a large curved cilium projecting beyond the body, by means of which they are believed to move, apparently taking its rise from the nucleus.

The form of luminosity due to Noctilucæ, although very striking, yet, owing to its softness, appears to be completely extinguished by moonlight, even when the moon is young. It was exhibited, only less marked, on the two following evenings, and on the third we left False Bay, a locality which has been remarked as very frequently exhibiting this beautiful phenomenon.

On the 7th of July, in lat. 28°, on the coast of China, only two days after the occurrence of this form of luminosity as first described, a heavy swell coming in from the south-west was met by a north-east wind, and the ship rolled tremendously. The sea was beautifully luminous, every wave breaking with a pale light, which was visible at a considerable distance, so that the whole sea was streaked with light; and again that peculiar phenomenon, of the ship sailing in a luminous sheath, was visible (see page 142). I mention this case, because it was one of the most striking instances of general luminosity which had come under my notice: it appeared to be compounded of the two forms I have already described.

The third form of luminosity to be described, consists of moon-shaped patches of steady white light, which I have found to be a very common phenomenon under certain circumstances. Next to the occurrence of sparks, and always accompanied by them, this form of luminosity is most frequently seen, and does not appear to be confined to any particular locality. I first observed it in the Mediterranean, on the first night on which the absence of the moon allowed it to be visible; and I have since found it to be no less frequent in the Red Sea, the Indian Ocean, the China Sea, and the Atlantic north and south of the Equator. It is most commonly visible in the wake of the ship, and consists of numerous round patches of light, closely resembling the appearance of white-hot shot, of various sizes, beneath the water at different depths. Sometimes, when deep down, they were pale, and of a whitish colour, with indistinct outline, and of large size; but when nearer the surface, they were smaller and more distinct, and assumed a pale greenish tinge. They usually remained visible for eight or ten seconds, but sometimes less. As these appearances were just such as might be presented by the umbrellas of large Medusæ, were such present and luminous, I was strongly inclined, at first, to attribute them to this cause; and the fact that on one occasion (about a week after I left England), I saw these moonlight patches in the Red Sea on the evening of a day on which the ship had passed through a shoal of Aureliæ, led me to attribute them to their presence. I supposed that the Aureliæ, struck by the screw, gave out their light under the excitation of the blow, and floated away luminous and dying. But I was forced to abandon this theory afterwards; for I have since many times watched for floating Medusæ before the light failed, and have not seen one for days and weeks together, and yet the moon-shaped patches have been as bright and as abundant as before. And again, when we have passed through a thick shoal of Medusæ towards evening, the luminous appearances have not been more marked than usual, but even less so. Moreover, having secured specimens of these Acalephs, they have not exhibited any luminosity whatever during the night. Although, however, I ceased to regard the Acalephæ as the source of the luminous patches in question, there can be no doubt that the great numbers which are always visible immediately under the stern, are due to the fact of the eddies of the ship exciting the emission of light in certain animals capable of exhibiting luminosity. Not, however, that similar appearances are never seen in other situations where they are unmolested, though I must say that in my experience this is rare. Thus, in the Indian Ocean, in lat. $12\frac{1}{2}°$ N. and long. $55°$ E. (bar. $30°$, therm. $82°$), among other appearances I noticed now and then a large luminous patch, with a roundish, irregular outline, pass by, emitting a pale and steady light, although out of the path of the ship; and on August 17th, being in a small boat on the coast of Borneo, in a strong breeze, after dark, I observed deep beneath the surface, and entirely apart from any influence of the oars, the appearance of large globes of white light, shining persistently and spontaneously in considerable numbers.

Although I long and constantly watched for the bodies which produced this remarkable and frequent luminous effect, I did so for a considerable time in vain. In vain I attempted to penetrate below the surface in search of any animals which could possibly originate the light. Although I could distinctly see the bottom of the ship's rudder, 19 feet deep, I could never detect a trace of any living thing within that depth by day; but no sooner did darkness supervene than they were often in the greatest abundance. It was on June 2nd, in lat. $28\frac{1}{2}°$ S., and long. $9°$ E., that I was at length witness of a circumstance which seemed to elucidate the question. Looking as usual over the stern, there were plenty of moon-shaped patches, accompanied by sparks unusually large and bright. The patches were remarkably persistent, and could be traced for nearly half a minute after the ship had passed. They were evidently at a considerable but varying distance below the surface of the water. When far down they appeared large and faint, and ill defined; but when nearer the surface they were smaller, brighter, and better defined. As I watched, one of the bright bodies whirled about by the eddy of the rudder, came absolutely to the surface, and exhibited a nearly rectangular form of great brilliancy, of a pale green colour, and, as far as I could judge, about six inches long by two broad. It instantly occurred to me that it was a Pyrosoma, and that this Ascidian was the usual cause of the phenomenon, the circular form of the patches being only an illusion produced by the diffusion of the light through a certain depth of water. I continued watching for a long time in the hope of seeing another; but although so good an opportunity did not occur again, many seemed to come near the surface, diminishing in size, but increasing in brilliancy as they did so—one in particular, very low down, suddenly gave out a dazzling brilliancy, which produced a momentary effulgence in the water all around.

I may mention that on a moonlight night, when the moon has been dimmed by fleecy clouds, I have been able distinctly to recognise these moonlight patches; but when the moon has shone out clearly they were no longer visible.

I have now to describe the fourth form of luminosity exhibited by marine animals, viz., momentary recurrent flashes of light. This form is nearly as commonly seen as the moon-shaped patches already described, which it very frequently, although perhaps not always, accompanies. If, however, the latter are well marked, the flashes are almost sure to be visible. I first observed them in the Indian Ocean, north of the line, and, since then, in the China seas and Atlantic. This appearance is very striking, but can only be seen under favourable circumstances, i.e., when the night is dark and the sea smooth. An indistinct transitory patch of light appears in the water as evanescent as a flash of lightning,—so rapidly does it come and go that it is difficult to fix the exact spot where it occurred. The brightness of the flash varies probably according to the depth of the animal producing it below the surface,—sometimes it is of considerable brilliancy, and sometimes so pale that it would not have been noticed but for its suddenness. The colour is always whitish, and the form of the flash round, brightest in the middle, and becoming indistinct at the circumference. I have on some occasions seen these flashes occur in such numbers and with such rapidity that it would have been impossible to count them; though, more commonly, they were comparatively few and far between.

But that which interested me most in these flashes of light was the fact that they always occurred at a distance from the path of the ship. Although I have seen them accompanying the moonshaped patches of light in the ship's wake, the places from which I could best observe the flashes were the forecastle or the gangways, when they could be seen in the smooth water several yards distant from the ship's side, and entirely uninterfered with by the ship's motion. This fact proved to me that there were spontaneous emissions of light by some animals deep below the surface, which voluntarily, and at intervals, gave out a bright coruscation. Moreover, although rarely, on following with the eye the spot where the flash appeared, it could be sometimes seen to re-appear further astern, as though the emission was recurrent at definite intervals, as has already been described in the case of the luminous beetles called fireflies at Singapore. I have also noticed on one occasion that the flash, instead of instantly disappearing, was followed by a faint glow, which vanished gradually; but whether this was an optical illusion of the retina or not I cannot be sure, though I believe not.

Whatever may be the animals which produce these luminous appearances, they must habitually swim at a considerable depth. I never was able to make out any definite outline of the light, which always appeared more or less spherical with faint edges, and sometimes the size and faintness of the flashes seemed to prove that the light must have been diffused by its passage through a great depth of water, which would also account for the whitish appearance of what is probably really greenish light. But I am strongly of opinion that the sources of the flashes and of the moon-shaped patches are identical—in the one case emitting their light spontaneously, and in the other under the excitation of the eddies produced by the ship, and especially by the screw-propeller when at work.

Before quitting the subject of these flashes I must not omit to mention that while at Singapore, having taken some small Medusæ in a towing-net in the straits, I placed them in a glass which stood by my bedside. In the night I observed them flashing brightly with instantaneous flashes, of the same character as those above referred to, although not the slightest shaking was applied to the bottle or irritation to the animals. So also the Noctilucæ of Singapore harbour, which I kept similarly in a bottle, flashed frequently with rapid and bright coruscations; and I am strongly disposed to believe that luminous marine animals, in health, and acting spontaneously without external irritation, always exhibit their luminosity in this manner; and that it is only when strong excitation is applied that they give out a steady but temporary glow.

There remains but one form of luminosity to be noticed, which, although I have never been so fortunate as to witness it myself, has been observed by others who have been longer at sea than I was. This is what has been called *milky sea*, an extraordinary phenomenon of rare occurrence. It has been described to me by one who has seen it, as a general luminous glow, not confined to the crests of ripples or to disturbed water, but occurring in perfectly calm weather, and looking as though the whole sea was composed of a whitish fluid, like milk, with no conspicuous bright spots or sparks. Such an appearance reflecting a faint light upwards, illuminates the ship, rendering every part of the rigging plainly visible; and inasmuch as it can only be seen in the absence of the moon, the contrast of the white glowing sea with the black sky produces an effect calculated to strike the observer with a kind of awe. Although I have met with persons who tell me they have not unfrequently seen this phenomenon, I am disposed to believe that it is extremely rare. One who has not really witnessed it at all might erroneously suppose that such an appearance as I have already alluded to as having twice occurred to me on the coast of China (when the ship seemed to be sailing in a luminous sheath), corresponded to the description of a milky sea; and in a small way perhaps it did so; and I considered it at the time as the nearest approach to that phenomenon I had ever observed. But the milky sea must be something *sui generis;* and I imagine it to be owing rather to a condition of the water under certain peculiar atmospheric or climatic influences, than to any extraordinary number of luminous animals in it. A circumstance which once occurred to me, seemed to throw some light upon the subject, and confirmed me in this opinion. Having put down the towing-net in the Formosa Channel, it collected a number of small Entomostraca, Megalopas, minute Medusæ, small Porpitæ, Pteropods, Annelids, Globigerinæ, &c., which I placed in a basin of sea-water; and not having finished my examination of them, they remained upon the table during the night. On stirring the water in the dark, the whole became faintly luminous, giving out a general glow, as if every particle were phosphorescent; the minute Crustacea, &c. appearing as bright spots in the luminous fluid. If the slimy substance in which, in some marine animals at least, a luminous property appears to reside, become diffused through the water, as it is probable it may be under certain combinations of conditions and circumstances, a general luminosity of the water may result, similar to that observed in milky sea, while the small sparks, doubtless in great abundance, would remain unnoticed in the universal glow, but would at the same time greatly enhance the general luminous effect.

There is a common idea that a southerly wind is peculiarly productive of luminosity in the sea; but according to my observations, this is an error. The winds most prevalent when luminosity has been well marked have been westerly, north-westerly, or even easterly—south being perhaps the least frequent; but probably the direction of the wind has no special influence in the matter. What the favourable conditions really are, it is as difficult to say as it is in the case of floating animals generally. I have seen remarkable exhibitions on one night followed by nearly absolute darkness on the next, the conditions of wind, weather, barometer, and thermometer, being inappreciably altered. Probably temperature is as important as any influence—the luminosity in the Mersey only occurs in summer. And in rounding the Cape of Good Hope during the winter season, scarcely any luminosity was exhibited during the month in which we were passing through the higher degrees of south latitude.

The animals which I have observed to possess luminous properties are not numerous. Many of the more minute animals taken in the towing-net appear to exhibit them, more particularly the small Crustacea (Entomostraca), and small Medusæ (Medusidæ). I have no reason to believe that the larger Medusæ (Lucernaridæ) as Aurelia, Pelagia, Rhizostoma, &c., exhibit any luminous powers, having kept specimens which have invariably failed to do so. Nor have I any experience of the Physophoridæ becoming luminous. I have never seen a luminous Porpita or Velella, and although on one occasion when magnificent specimens of Portuguese men-of-war had been floating by all day, my attention was directed to shining spots at night, under the supposition that they were luminous Physaliæ, I merely replied by pointing to a bucket containing one of these animals, but which was perfectly dark. I have seen a large prawn give out light after death, and a fresh squid was illuminated at night with an irregular glow of whitish light, which remained unaltered as I passed my finger over the surface. Nor do I believe many of the stories of luminous fish, inasmuch as a fish rapidly swimming in a fluid abounding in minute luminous points, as the sea sometimes does, would present an effect which an uninformed or inaccurate observer would readily mistake as proceeding from the fish, itself, instead of from luminous points which it disturbed in its passage.

Few luminous marine animals have received greater attention than the Noctilucæ, and from them we may perhaps gather some indications of the seat and nature of this wonderful appearance. M. De Quatrefages concludes, as the result of his experiments with these Protozoa, that they retain their luminosity so long as they continue to possess *organic contractility.* So also in luminous Annelids,—or perhaps, as better expressed, because the subjects were larger and more highly organised, " in the great majority of cases the light manifests itself in scintillations along the course of the muscles alone, and only during their contraction." The light is entirely unaccompanied by heat, nor is there anything analogous to a combustion, either active or slow, of a chemical nature. So also Kölliker, in his examination of the luminous property of the glow-worm (Lampyris), came to the conclusion that there was neither combustion nor phosphorus in the case; but that it was the product of a nervous apparatus, and dependent upon the will of the animal.

Ever since, many years ago, I became acquainted with Mr. Grove's researches upon the Correlation of Physical Forces, I have looked upon that ingenious theory as the rational explanation of animal luminosity. Light, heat, electricity, magnetism, motion, and chemical force, are all interchangeable, and each may manifest itself in the form of the other; but although these are called the *physical* forces, who can say that they are not *organic* forces also? One of them, which long since would have been regarded as eminently inorganic, is now fully recognised as an organic force, produced by vital organs, and regulated by the will of the animal exhibiting it. I allude, of course, to electricity, an agent which is possessed by several fishes, and we know not by what other animals,—a force which is produced directly through the agency of nervous power, for the regulation of which a special cerebral lobe is recognised. If this nerve force or vitality can display itself in the form of electricity, why should it not do so also in the form of light? In the more highly organised luminous animals, as in Lampyris (the glow-worm), in which nervous centres exist, there is a special organ for the development of light, doubtless regulated by some part of the nervous system. In others, the contractility of muscular tissue or of sarcode substance, which contractility is itself a vital act, seems sufficient to produce the phenomenon in question. In animals which have no definite nervous system, we can scarcely predicate the existence of a will; and, therefore, while the glow-worm and many of the higher marine luminous animals probably exercise a control over the functions of their light-giving organs,—in the Noctilucæ, and such lowly organisms, any external excitement which produces a temporary contraction, is at the same time sufficient to exhibit its correlative accompaniment, light.

There does certainly appear to be a phase of luminousness which is scarcely of this character, and which takes place both on land and in water. I refer to a luminous coat external to the animal, as in the case of the trail of light left by the little Scolopendra electrica on the ground, and the luminous mucus which exists in certain Medusæ, and, it is said, Pholads, or boring Mollusks, and which retains its properties apart from the animals. But whatever may be the explanation of these phenomena, there can be no doubt that they are no more due to combustion or to phosphorus than the appearances exhibited by the specialised organs of the glow-worm, but may more probably belong to another group of facts, that, namely, which includes the luminous appearances presented by certain plants, in which we cannot call to our assistance either nerve force or vital contractility.

Organic luminousness exhibits itself with a wonderful range over the animal kingdom; and if we were called upon to specify in what classes of animals it has been observed, we should prefer to make a converse statement, to the effect that, with the sole exception of birds, every class of animals, under certain circumstances and conditions, has been proved to be capable of giving out light.

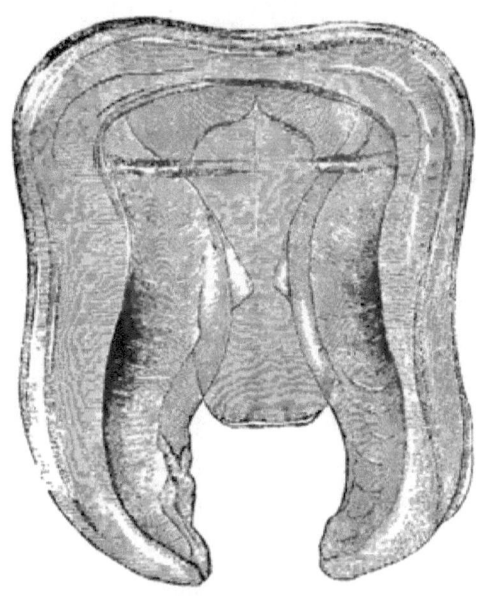

Berobid Ciliograde, from the Atlantic.

CHAPTER XXIV.

THE VOYAGE HOME.

Storm at Hong Kong—Loss of the "Osprey"—Sea-birds at the Cape—Simon's Bay—Cormorants—Botany of the Cape—Physical Features of Table Bay—Cape Town—Marine Animals of Simon's Bay—Coast of St. Helena—James' Town—Napoleon's Tomb—Ascension—General Features—Crater—Vegetation—Insects—"Wide-awake Fair"—Boldness of the Birds—Turtle Ponds—Varieties of Turtle—Western Isles—Pico—Fayal—Villa de Horta—Character of Vegetation—Spithead—Conclusion.

On the 24th March I embarked on board H.M.S. "Scylla," 21 guns, for the long voyage from Hong Kong to England, a distance of 13,000 miles. The time of year promised favourable weather and fair winds, and the promise was, on the whole, performed. We began by escaping a violent storm at Hong Kong, for scarcely had we passed Green Island when the sky became extraordinarily dark and gloomy, and one of the most tremendous squalls which I have ever witnessed raged behind us. The blackness which closed in round three sides of us made it so dark as to give rise to the general remark that it was like an eclipse. The uniform and unbroken mass of cloud near the horizon assumed a ghastly green tint, which was equally unusual and extraordinary; and the heavy black clouds rolled towards us, curling over and over, and hanging down in murky festoons which threatened to form into great waterspouts. For a long time I watched these effects, untouched by a drop of rain; but at length it fell, though not heavily; and with it came the wind, which lashed the sea into foam, while the most terrific lightning flashed through the green-black sky. But we had fair weather ahead, and only the skirts of the storm reached us, while the full force of its violence was felt at Canton and Hong Kong. At Canton tremendous hailstones, as large as pigeons' eggs, fell and did considerable mischief, as well as injury to the population; while at Hong Kong the storm raged all day with unusual severity.

After a delightful passage we reached Singapore on April 3rd, and once more continued our journey on the 11th. As H.M.S. "Osprey" was to leave Singapore a few days after us, and would probably reach the Cape before we left, it was arranged she should bring us our next mails; and we had, moreover, exchanged some officers with that ship. But the "Osprey" was destined never to reach Simon's Bay: she was wrecked and lost off Cape Agulhas, the first intimation of which reached us after we had weighed anchor and were standing out from St. Helena, when the Cape mail arrived, and we were all set speculating by a signal from the station ship, which we were just within distance to read—"*Osprey*" *lost; all hands saved.*

The voyage across the Indian Ocean, from Java Head to the South African coast, was long and uneventful. As we approached the coast large numbers of sea birds almost constantly accompanied the ship, of which perhaps the most common was the fork-tailed petrel (Thalassidroma Leachii). The first time I observed them was 1300 miles west of Java Head, where the only intervening land was the Keeling Islands. They flew about the ship's wake, skimming over the crests of the waves, and evidently quite at ease; every now and then putting out their legs as though touching the surface, or running along the water. Night and day they followed the ship for weeks together, never appearing to rest. But whence they come, where they live, how they sleep or rest, is a mystery. If on the water, why are they not snapped up by predaceous fish? That some of these birds are of nocturnal habits is proved by the fact that I have seen, when 24 hours' sail from Ascension, a bird hovering over the ship in the moonlight, sailing to and fro across the moon's disc, for some hours, and that when no birds had been observed about the ship for some days. Tropic birds (Phaethon) also made their appearance in long. $62°$ E., and lat. $23\frac{1}{2}°$ S., when we were 17 or 18 days distant from land in either direction. But when near the South African coast the fork-tailed petrels were accompanied by the sooty petrel, or Cape hens (Puffinus major), Cape pigeons (Daption capense), mollymawks (Diomedea chlororhynchus), whale-birds, and even gannet (Sula capensis), which last I observed at least 70 miles from land. When, however, we were blown off the coast by a north-wester, the fork-tailed petrels, and mollymawks, or yellow-billed albatross, were the only birds which seemed at home in the gale. The latter are singularly graceful in their flight. They swim well and rapidly; and when leaving the water assist themselves to rise by their feet, running quickly for some distance along the surface until they are fairly above the water. How they propel themselves in the air is difficult to understand; for they scarcely ever flap their wings, but sail gracefully along, swaying from side to side, sometimes skimming the water so closely that the point of one wing dips into it, then rising up like a boomerang into the air,—then descending again, and flying with the wind, or against it, apparently with equal facility. Now and then, but seldom, they give two rapid flaps with their wings, but to see this they must be watched.

After having been twice repulsed, we entered False Bay on May 23rd, and anchored off Simon's Town, where we were consoled for not having touched at the Mauritius, by the news that that island was scourged by a terrible plague which had been brought, it was said, by some Chinese coolies, and had spread with frightful rapidity, decimating the inhabitants, more especially however the native population, and driving numbers away to seek for safety in more healthy places. Had we called there we could have held no communication with the shore, and should have gone out of our way for no purpose.

Simon's Bay is a sheltered corner of False Bay, one of the largest and deepest of the South African bays, which has the narrow range constituting the Cape of Good Hope between it and the Atlantic. Ships passing round from the East have occasionally entered it, believing that they have rounded the Cape, and hence its name. This bay is, of course, greatly exposed to southern gales, the rollers from which run fairly into it; but as the north-west corner is hollowed out, it forms a well protected harbour, around which a naval establishment and small town have arisen.

Here all Her Majesty's vessels touch in rounding the Cape, Table Bay being too exposed a situation, with no shelter from west and north-westerly gales, though in this latter place all merchant ships anchor. Simon's Town, as it is called, is a cluster of white houses at the base of the lofty and barren hills which connect the Cape of Good Hope with Cape Town. The shores are piled with boulders of granite, and more than one large tabular mass appears above the water, affording resting-places to thousands of cormorants, which during the day make periodical excursions to the sandy beaches at the head of False Bay, streaming in long black lines along the surface of the water, and at midday leaving without a single occupant those rocks which at mornings and evenings are blackened with their numbers. They were of the common species—(Phalacrocorax carbo). It was curious to watch these birds feeding here and there upon the shore upon substances thrown up by the tide. Standing in the water just where the waves broke, they would have been tumbled over by every advancing billow, but, watching the critical moment, they would rapidly dive under the wave and thus avoid it, and then proceed with their search till the next came, when they would repeat the process.

Such numbers of fish-eating birds (and besides these there were plenty of gannets and terns) argue large quantities of fish, and Simon's Bay has the reputation of abounding with them. I myself saw, when looking over the bay from a slight elevation, at least half a dozen large shoals at the same time, splashing about and disturbing the water like so many cat's-paws upon the surface. One of these shoals, which I observed close under the ship, consisted of myriads of fishes, averaging from two to eight inches long.

The first thing which strikes the botanist on landing at

Simon's Town is the number of Opuntias (Cacti) which grow upon the rocks, a circumstance at least remarkable when it is remembered that the Cactaceæ are an American order, unknown in Africa. I believe, however, that having been imported from America, they have found a suitable habitat upon rocks of the Cape, and have readily become naturalized there, and spread throughout the colony. The representative orders, Crassulaceæ and Euphorbiaceæ, however, abound everywhere. Great aloes, also, with fruiting stems 20 feet high—trees of Oleander and Casuarina, and other remarkable vegetable forms, strike the eye as novel and interesting: and among the abundant verdure at the foot of the hills elegant herbaceous Amaryllids and Cape heaths (Ericaceæ) meet the eye in every direction.

False Bay, seen from the Cape road which runs southward from Simon's Bay, has the appearance of a vast lake, closed in on the opposite side by a long line of craggy peaks, which are misty and indistinct from distance, and which bound the eastern side of the bay. Instead, however, of closing round the north side, they continue to run on inland with a northerly course as far as the eye can reach, while the precipitous and rugged mountains on the western side, which arise directly from the sea at the Cape of Good Hope, also run northward and terminate in Table Mountain, which slopes immediately upon Table Bay. Between these the waters of False Bay wash upon a sloping sandy beach, and a tolerably level plain extends between the two ranges for a considerable distance, as though it were at one time deeper and of considerably greater extent than it is at present. An indifferent road skirts the bases of the hills on the western side towards Cape Town, which is, in many places, diverted on to the sea-beach, and at high-tide the wheels of the conveyance are washed up to the axles by the advancing waves. Where, however, a legitimate road exists, it runs through numerous fishing stations, and beside cottages and gardens betokening a well-populated country. One suggestive circumstance I remarked in passing the head of the bay, viz. the frequency with which the fences were formed of, or replaced by, the ribs and other bones of whales, proving how commonly these animals were washed up on the beach by the southerly winds.

Farther on, the numerous villas, with plantations and gardens, indicate the proximity to Cape Town, the higher classes of which almost universally live at a distance from town, the convenience of a railway from Cape Town to Wynberg favouring this practice. This, in fact, may be regarded as the principal suburb of the capital, although situated at a distance of seven or eight miles from it. It seems a pity that the railway is not further extended to Simon's Town; but there appears to be no chance of such an extension, for the colony is unfortunately not at present in so flourishing a condition as to be able to lay out so large a sum of money. Indeed, owing to the much superior anchorage of Simon's Bay, it may be anticipated that many ships now anchoring in Table Bay would prefer the former, if cargo could be readily transported across to Cape Town, so that the construction of a railway would possibly tend to injure the latter place.

Cape Town is handsomely built: the streets wide and the shops good. Perhaps the finest public building is the South African Museum, superintended by Mr. E. L. Layard. I did not observe much in it of special interest, except some flint implements, which Mr. Layard pointed out to me as having been recently found in the colony, and which have the unmistakeable impress of relationship to those found in Europe, certainly a very remarkable fact, and wonderfully extending the geographical area of those early inhabitants of the earth whose first traces have been so ably followed up in Great Britain, France, Switzerland, and Denmark.

Attached to the Museum is an admirable library and reading-room, which includes the munificent gift of Sir George Grey, the late Governor, and which was rich in valuable MSS. and early editions of works dear to the bibliophile.

Mr. M'Gibbon, the curator of the Botanic Gardens, kindly accompanied me over that establishment, which adjoins the Museum. I hoped to have found a collection of Cape plants in it, but was disappointed, and but few plants had names attached to them. The Colonial Government grants the sum of £250 per annum for its support, which, it must be allowed, is small enough, and for which the respectable condition of the garden was ample return.

The sea-shore at Simon's Bay is strewed with boulders of grey granite, much exposed, but affording some sheltered crannies, in which were beautiful natural aquaria, containing a number of Actiniæ of a crimson colour, closely resembling our A. mesembryanthemum, as well as others of a white and buff colour. The dominant mollusk was certainly Patella (Limpet); not only did numerous forms of Patellæ strew the sand, but large and handsome ones adhered to the rocks, overgrown with seaweed, and looking like little moving pastures. In one of these large limpets (Patella oculus) which I removed, I found a very pretty mottled Planaria ensconced under the mantle. Next to Limpets, Trochi were most common. Two Echinoderms, both probably new, rewarded my search. One of them was an Asterina, most beautifully variegated with bright red, white, and blue, but extremely variable in pattern, so that no two of them were alike. The arrangement of the colour was in spots or papillæ, precisely resembling the effect of Berlin-wool work, each papilla being of some definite colour, and adjacent papillæ being often strongly contrasted. It was in considerable profusion on the rocks leading to the Cape of Good Hope. The other Echinoderm was a small Echinus, also extremely variable in colouring. There were three distinct varieties,—the spines of one of a rich scarlet, with an undergrowth of small olive-green spines, and tentacles of metallic blue; the second variety had deep violet spines and brown tentacles; and the third, pale reddish or buff, the most common form.

We left Simon's Bay on 28th May, and on the 8th of June cast anchor at James Town, St. Helena. The approach to this remarkable island is very interesting; and two pointed peaks in the east and west end of it respectively, having steep escarpments towards the sea, and smooth, gentle descents on the landward side, bear witness to its volcanic origin, and have every appearance of being remains of two sides of a great crater, whose other sides had been washed away or demolished by the inroads of the sea through long continued ages. Very bold outlines and peaks, with rich contrasts of light and shade, were developed on a nearer approach; and when quite close, the points of interest in the rugged coast greatly multiplied. Sharp, serrated peaks,—rows of basaltic crags, with the characteristic sloping talus of *débris* at their bases,—gigantic faces of rock, with thin dykes traversing them irregularly from top to bottom, at right angles to their cleavage,—all formed a fine study, and but too rapidly passed before the eye.

One magnificent cliff of black basalt was particularly striking as we rounded the eastern corner of the island, having numerous overhanging ledges and ridges of various lengths running along its face. In two or three spots forts had been built, connected by covered ways, while at the base of the rock large caverns were excavated, into which the waves dashed, casting the spray into their arched depths.

James Town is situated in a narrow gorge which winds between two lofty barren hills, and extends fully a mile inland, where the gorge terminates in a cul-de-sac, at the bottom of which is situated "The Briars," a pretty cluster of buildings, where Napoleon took up his residence upon his first arrival in the island. In the valley are gardens with cocoa nuts and bananas, which form a pleasant relief to the great barren wall of rock everywhere overhanging it. Pursuing the path which winds up the left side of the valley, the rocks are seen to be covered with the prickly pear (Opuntia), which I had remarked as a naturalized plant at the Cape, and is here in great abundance, having the reputation of being an importation from the West Indies. Mingled with it were great numbers of scarlet geraniums (Pelargonium), now in full bloom. White Daturæ (May-apple) and other plants grew by the road-side, and at the top of the ridge were pretty and shady woods of willows, with bushes of Buddlea, the open spots being covered with real English gorse (Ulex europæa) full of yellow blossoms, and growing in a good rich soil.

From this hill could be seen those points which render the island historically interesting in connection with Napoleon Bonaparte. Looking across a most desolate valley, the bottom of which was deep and dark, the cluster of buildings constituting Longwood might be seen peeping out of a few trees which crowned the crest of the opposite ridge; and down in the valley where it narrowed on the right, and became more pleasing and somewhat verdant, was the spot where the great Emperor found his last small but sufficient empire. His bones are no longer there; but the spot is still venerated by those who hold dear the traditions of the empire, and who reverence the name of the man whose restless ambition desolated thousands of homes, and covered France with glory. A massive square slab of stone, surrounded by an iron railing, within which are planted scarlet geraniums, and over which hangs a stunted weeping willow, is the true description of this "last scene of all of his eventful history." The enclosure is small, hilly on three sides, and planted around with trees, among which the sombre tints of the funereal cypress are conspicuous.

The island of St. Helena appears to abound with beautiful views and varied scenery, and I very much regretted having to leave it without a fuller exploration; but we weighed anchor next day, and, with a fair wind, stood away for Ascension, and once more cast anchor off George Town on the 13th of June.

The volcanic island of Ascension, as approached from the sea, has not so striking an appearance as St. Helena; but owing to its peculiar and predominant rufous colour, and desolate aspect, it is not a little remarkable. Sloping rocks of the roughest lava, broken here and there by sandy bays, stretch along the shore, and the island consists of an irregular series of conical hills of various heights, above which towers Green Mountain, 2800 feet high, whose summit is crowned with trees and green fields, and offers a strong contrast to the other hills, which are reddish or brown, according to the colour of the ashes and cinders of which they are composed.

The settlement of George Town is entirely naval in its character, being formed of a number of departmental officers, and of marines, who are all borne on the books of H.M.S. "Flora," 40 guns, which lies off this place, and whose captain is styled the "Captain of the Island." Everything is conducted with the strictest reference to naval discipline, and the island is nothing more nor less than a ship ashore. The landing-place is very indifferent, mere steps cut in the rock, and therefore entirely inaccessible in bad weather. It is well known that the great waves of the Atlantic often set in upon the rock in the form of *rollers*, even in fine weather, and it can never be predicted when they may make their appearance; but whenever they do so all communication between the ships and the shore is cut off, except by signal. It is one of the duties of the master of the "Flora" to direct a flag to be hoisted on the signal-hill when this state of things occurs, and that is pretty frequently.

As our stay was to be limited to one day I was thankful that the weather was calm and the sea permitted us to land; and having done so, I bent my steps in the direction of South-West Bay, with the intention of visiting "Wide-awake Fair," and at the same time exploring some of the geological features of this remarkable island. The whole of Ascension is an erupted mass, the antiquity of which can only be judged of by the worn condition of its surface; but it is entirely the product of a once active, but long since extinct, volcano. Green Mountain, the culminating point, is probably the parent cone, around which a great number of secondary cones and craters are clustered, the rough trachytic lavas of which run sloping to the beach round the greater part of the island.

One or two tolerable roads have been formed, which greatly

save the labour of walking in a country where the surface of the ground is heaped with rough and sharp-pointed cinders, which look like the product of a myriad furnaces, and to which the "black country" of Staffordshire is a trifle. From these arise conical hills of a reddish colour, covered with fine ashes, which crackle under the feet, and from out of which peep the rounded overhanging ledges formed by molten lava. Down these hills streams of water have poured during the brief and uncertain wet seasons, forming water-courses which run between the rounded knolls, which look like *roches moutonnées* at the base, and intersect the lava-fields down to the beach. For rain falls occasionally on the island, though unfrequently; and on the day of our arrival it was wet: it rained all night, and next morning Green Mountain was enveloped in cloud. Other hills are hollow and crateriform, the sides formed of loose masses of slag or clinker of various sizes. Up one of these I clambered, and found the interior deep and cup-shaped, but incomplete on one side, the bottom being a small level deposit of mud and sand, produced by the washings of the cinders in wet weather; among these cinders I found several fragments of exploded volcanic bombs, such as are described and figured by Mr. Darwin in his notice of the island.

From this elevation the view was most striking: a deep and broad rocky valley in the fore ground, covered with screaming sea-fowl, beyond which arose an irregular series of naked and desolate conical hills piled one above another in chaotic confusion, but surmounted by the verdant and fertile heights of Green Mountain, upon which may be descried trees, meadows and pastures, like the Delectable Mountains seen afar off by the pilgrims.

It must not be supposed, however, that the surface of the

island is absolutely without vegetation. The cinders in many places are incrusted with white and gray lichens (Parmelia and Roscella). Some are overgrown with more luxuriant species, as Physcia cæsia, and I also observed a deep-green incrusting lichen on the sea-shore. Many spots, also, in the water-courses, are quite cheerful with patches of bright green, and several flowers spring up here and there which have escaped from the gardens on Green Mountain. I was informed that some person had been in the habit of scattering seeds over various parts of the island. I noticed two species of grasses, a Sonchus, an Aster with scented leaves, &c. The most common plants, however, were the castor-oil (Ricinus), a very handsome yellow poppy with prickly white-veined leaves, and a large-flowered plant (Vinca rosea) which is known on the island as the Madagascar Rose, and is reported to have been imported from thence.

Among this vegetation a few insects occur : large red-winged locusts fly about among the rocks, and a fat black cricket is common—I also saw a pale brown one, but could not catch it. A little moth, very prettily marked, is common wherever a certain succulent plant occurred, and flew about among the rocks, settling for a moment and then taking wing again, unless it happened to get in the shelter of a crevice in the honeycomb of a cinder, where it seemed to consider itself safe.* A somewhat larger pale brown moth I also noticed from time to time; but it flew rapidly and was aided by a strong breeze which was blowing, and appears usually to be blowing, over the island. Besides

* My friend, Mr. Stainton, informs me that this little moth is Hymenia recurvalis, of Fabricius, and that the British Museum possesses specimens of it from Jamaica, Sierra Leone, Ceylon, Bagdad, India, China, Australia, and New Zealand. To these localities must now be added Ascension, so that this little feeble insect is literally cosmopolitan.

these insects I saw carrion flies upon the rocks, a hunting spider, and numerous small carrion beetles (Dermestes) in situations to be presently mentioned.

The lava and cinders in the neighbourhood of South-West Bay are whitened here and there by the dung of sea-birds; but the extraordinary scene of the breeding-place of the terns, or wide-awakes, and called "Wide-awake Fair," is a long valley situated about half a mile from the sea in the south-eastern part of the island. The approach to this valley is indicated by an overpowering odour arising from their deposits, which, however, do not accumulate as in some guano islands. Seen from the hill above, this valley looks as though a light fall of snow had partially whitened it; but in no place was there any appreciable depth of deposit. The birds themselves are in immense numbers, hovering over the valley, screaming and making various discordant noises, which, heard at a distance, sound like the murmur of a vast crowd. They are elegant and graceful birds, glossy black above and snowy white beneath, with white foreheads, straight compressed beaks and long forked tails: they measured 2 ft. 6 in. from tip to tip of the wings, which are long and pointed. As soon as a visitor makes his appearance among the nests, numberless birds arise screaming in the air, and form a complete canopy over his head; some, bolder than the rest, fly so close that it is the easiest thing in the world to knock them down with a stick, and it is even necessary to strike at them occasionally and give them a slight tap to admonish them not to use their bills against one's face. Meantime crowds of little ones, of all ages and sizes, some covered with a grey down and others almost fully fledged, run hither and thither, tumbling over the stones in their hurry to escape from the intruder. Here a chick has but just broken the egg, and the parent bird is nestling over it, and does not leave it until you arrive so close that you could stretch out your arm and take it up. Eggs lay scattered all over the place, deposited in little hollows in the sand, about as large as the palm of the hand, which is all the nest that the "wide-awake" considers necessary; and in several of the rocky crevices in which these eggs were deposited the skeleton or half-decayed body of an adult bird, but more frequently a young one, upon which a number of carrion beetles were busy, showed where it had died and rotted beside the nest.

At the particular season at which I visited this singular spot, the birds were in every stage of growth, from the newly-hatched chick to the bird with first year's plumage, flying with the rest. Eggs also were abundant, but never more than one in the same nest; and although the parent bird was in some cases sitting upon fresh or half-hatched ones, in a great many instances the eggs were cracked, and either rotten or dried up. Many that I picked up felt light and empty, although scarcely injured, and others which I broke contained carrion beetles or their grubs. The eggs were very variously marked, and had not a little variety of form: the common appearance of them was round at one end and pointed at the other, about the size of a plover's egg, and in colour a whitish ground, blotched with faint purplish and distinct rich brown blotches, which often formed a ring round the larger end; but some which I noticed were long and pointed at both ends, and without blotches, but speckled with small purplish and brown spots. There was no other kind of bird, however, visible in the whole valley.

It would be easy for any person to fill a sack with adult birds, although he possessed no other weapon than a stick;

and too many of the visitors are not content without maiming a number in mere wantonness; so that the poor birds can hardly be said to dwell unmolested; nevertheless, as long at least as they have nests and eggs to look after, they evince what I should characterize as boldness rather than tameness. I should consider the Solan geese on the Bass Rock as tamer than the "wide-awakes" of Ascension.

Before leaving the island I visited the turtle-ponds, where these animals are kept in store; for Ascension, barren and desolate as it is, has yet one product in which it is not exceeded by any part of the world, viz., turtle. The sandy bays of the island are visited by great numbers of these unwieldy and valuable reptiles, which, entirely marine and oceanic in their habits, visit the shore solely for the deposition of their eggs, and are secured on these occasions by being cut off from their retreat to the sea and turned over on their backs, and then conveyed at leisure to the reservoirs provided for their reception. The sandy shore adjoining George Town, I was informed, is no longer so rich and profitable a beach as it once was, the reason probably being that turtle, like birds of passage, return again and again to the same spot to deposit their eggs; and on this beach, as being most accessible, the greatest number of turtle have been turned, so that but few visit it at present. No one but the government authorities is allowed to interfere with this source of emolument, and the turtle form a staple article of food upon the island, being served out twice a week; but the animals are sent to persons in authority in England, and are supplied to merchant ships at the rate of £2 10s. each. The season was just over when I visited Ascension, and the turtle-ponds contained eighty-two animals. These ponds, two in number, were on the sea-beach, each 50 or 60 feet square, and three or four feet deep, and the sea is allowed to wash into them through two grated channels. All the turtle, however, were in one of these enclosures, and could be seen swimming about, ever and anon raising their stupid-looking heads above the surface and snorting out a jet of water. They seemed to crowd together in one corner, where each wave as it broke sent a rush of fresh sea-water into the pond. Numerous small fishes and crabs swam about them unmolested; but on inquiry I learned that they are never fed, although they are not unfrequently kept in the reservoirs for a year or more after capture. They were very variously marked, some with large black spots, others with indistinct radiating streaks upon the plates, and several had a large white patch in the middle of the carapace. One in particular was conspicuous from its very peculiar form. Instead of being gently rounded as usual, the carapace was high and terminated in a ridge, which, as it swam about, was elevated fully six inches above the water—a conformation which it appears occasionally, although rarely, occurs. While I was watching them, preparations were made for getting one out of the pond. A negro walked into the midst of them, and having selected one, he tied a cord round one of the anterior fins, by which it was pulled by several other negroes out of the pond by main force, and laid upon its back on a small four-wheeled carriage prepared for it, in which helpless position it was dragged away without a struggle.

On the lava rock adjacent, where the waves break with great violence, numbers of beautifully coloured crabs (Grapsi, n. s.) ran actively about; the pools abounded with large purple-spined Echini, ensconced in round hollows, and beautiful azure and banded rock fish; but the only seaweed I observed was the cosmopolitan peacock's-tail (Padina pavonia). My exploration, however, was necessarily brief, as I was obliged forthwith to rejoin the ship.

A long succession of calms, with occasional light winds from the north-east, carried us so far west, and consumed so much coal, that it was determined to make for the Azores, and put into the port of Horta, in the island of Fayal, for fresh supplies. Accordingly, just as we had left the Sargasso Sea behind, we came in sight of the islands on July 11th. The first sight of the western isles from the south is very remarkable; the island immediately east of Fayal, called Pico, towers up to the height of 7600 feet, and the pointed apex appeared just above a belt of cloud, and seemed to reach the very skies, its apparent height being greatly magnified by this circumstance. All the morning, as we neared it, the clouds varied but little, and the black top seemed almost to overhang the ship. Nor was the island of Fayal less interesting, although in a different way. The whole southern side, gently sloping from a long ridge towards the shore, was most beautifully cultivated, and mapped out into yellow fields, interspersed with green patches high up the hill side; while numerous white cottages were dotted over the landscape. Many parallel ravines running down the slope afforded sheltered spots, in which white houses were clustered; and the whole formed a delightful picture of fertility and repose, which gradually opened up new points of beauty as we came nearer.

Standing on the forecastle admiring this picture, I could turn to the calm sea in which clusters of Salpæ were swimming, accompanied by various species of jelly-fish, while flocks of sea birds sat here and there upon the water, and porpoises rolled about merrily. Presently the trenchant fin of a shark appeared right ahead; and leaning over the bows,

I watched him swimming hither and thither, until he was nearly under the cut-water, when, to my delight, I observed it to be a large shark of the hammer-headed species (Zygæna malleus), twelve or fourteen feet long. While he was fully in view he suddenly darted off out of danger; but his fin could be seen for some time after on the quarter.

As we rounded the east side of Fayal to enter Horta Bay we passed two great shapeless rocks, one of a dark brown and the other red, which showed the volcanic structure of the island; for all this fertility and verdure cover an extinct volcano, and even yet shocks of earthquake are not unfrequent. It was like the cloven hoof peeping out from under a gorgeous robe; but even one of these rocks was terraced with vines on the landward side.

The Villa de Horta is a charmingly situated place, and looks extremely pretty from any point, set as it is against a background of highly cultivated fields and hills; while the majestic mountain of Pico, on the other side of the bay, forms a fine object from the landward. When disencumbered of clouds, however, it did not appear nearly so lofty as when we first saw it. Small craters may be seen upon its sloping sides, as well as some cultivation and a few white houses; but it is said that the great cone occasionally smokes. Only a week or two before my visit a small island was thrown up in the sea near St. Michael by volcanic agency.*

---

* During the past year the Azores have been the theatre of unusual volcanic excitement, extending from December 1866 to August 1867. On May 25th last, between half-past two and midnight, there were experienced no less than fifty-seven distinct shocks of earthquake. Five days later the ground was in constant motion—and on the 1st of June there was a violent earthquake and volcanic eruption, the day after which the sea was covered with a layer of sulphur—the water appeared to boil, and jets were thrown up. On the 4th of June sulphuretted hydrogen fumes were given out; there was an earthquake

The streets of Horta were quiet and very hot, and the people of all classes (Portuguese) extremely polite. Here we purchased fresh beef at fourpence halfpenny per pound, and small fowls at four shillings the dozen. Priests and nuns, the latter with enormous heavy black hoods, which gave them an extraordinary appearance, abound; and, if report spoke correctly, the people stand in some awe of the former. The Freemasons, particularly, were forced to hold their meetings by stealth, owing to the anxiety shown by the priests to become acquainted with the masonic mysteries. Outside the town the walks are not so pleasant as appeared from the sea; for the roads are all sunk between lofty rough stone walls, and there is no shelter whatever from the rays of the sun. The chief crop seemed to be rye, which was ripening for the harvest.

I was much struck in my walk in Fayal with the great resemblance which its natural productions bore to those of this country. The road-side vegetation, which was all I could observe, seemed quite familiar to me, and I gathered the following common British flowers as I went along, viz., Geranium molle, Lapsana communis, Sisymbrium officinale, Verbena officinalis, Cotyledon umbilicus, Veronica officinalis, Trifolium album, Malva rotundifolia, and Hordeum sylvaticum. The common white butterfly (Pontia) crossed my path from time to time; and I saw, and heard sing, the chaffinch (Fringilla cœlebs) for the first time since I left England.

As we quitted Fayal the island of St. George had the aspect of a long rugged rock, with steep sides, cut into numerous ravines, all having an uniform neutral tint, dashed here and there with green. Terceira lay in the distance among the clouds of the eastern horizon; and Gloriosa we

on the 27th, and a slight oscillation on the 17th of August, since which it appears that the volcanic forces have been more or less at rest.

passed near enough to see that it was cultivated and mapped into fields, and had a large town on the west side, near which rose a *sacro monte* of considerable size, whose white chapels glistened in the sun.

A favourable breeze brought us in a few more days into the Channel, and on the 21st July we dropped anchor at Spithead.

In conclusion, and by way of retrospect, I may safely say that any one who will undertake such a voyage as that whose incidents I have recounted in the foregoing pages, will find himself amply repaid by the stores of information which he will insensibly but surely acquire by the constant observation of the phenomena around him; and he cannot fail at once to enrich his own mind, and to benefit science, if he will only faithfully use the opportunities which fall to his lot. These opportunities of course will vary with circumstances, and are unfortunately, under the most favourable conditions, not all that could be desired or wished; nevertheless they will sometimes occur, even in the most adverse cases, and so much is to be learned of marine animal life within the tropics, that the most striking and novel facts come to light when least expected or looked for. Whenever the dredge can be used, a rich harvest is almost sure to result—proportionate to the rarity of the opportunity, which in my experience was very great. But much may always be done by a diligent investigator upon the shore, or by wading in shallow places. And even when at sea, one need never be idle, for there is abundant occupation for the microscope and the pencil in the contents of the towing-net, or in the numerous organisms which may be observed in occasional calms. The great drawback in a ship of war is the necessity for using steam when no longer propelled by the wind, and the difficulty of procuring a boat to be lowered when anything unusual or novel is seen floating, and which might by this means be easily procured. On my way home I often devoutly wished I was in a sailing vessel, that I might have revelled in the wonderful richness and variety of animal forms which nearly a thousand miles of calms afforded; and doubtless greater opportunities would be enjoyed in a cruise with an intelligent merchant captain, than in a man-of-war —especially if in the latter case one is hampered by the crotchets and caprices of an unsympathetic commander. The chief drawback to travelling in merchant vessels would be of course their more limited range, and their avoidance of intermediate ports—circumstances which, however, might or might not be hostile to researches of this nature. They would, however, naturally avoid reefs as they would poison; and interesting land journeys would seldom be possible to those using them.

I cannot help feeling great regret that the wonderful advantages which fall to the lot of many of our naval officers are so totally lost. They have their duties on board ship to perform, it is true, but that some of them should not have learned to relieve the dull and unendurable monotony of sea life by such studies is to me unaccountable. The medical officers especially, whose education would most fit them for these pursuits, and who have by far the most leisure at their disposal, might be expected to follow them with no less of advantage to themselves than of benefit to science; but it is only one in a thousand who troubles himself to observe what passes around him, or makes any exertion to share in the reputation acquired by a few of their fellow-surgeons, such as an Adams, or a Macdonald.

Indeed there seems to me to be more hope of valuable materials being accumulated by the better class of merchant-skippers than by any branch of the naval service; and I am acquainted with several merchant-captains who bring home at the conclusion of every voyage industriously-formed collections and intelligently-written observations, which are of much interest, and are yearly becoming of more importance. I have been at some pains to bring about this end, and have to some extent succeeded, though time alone can evolve a more complete and general practical result. The Admiralty has long since issued a useful scientific manual for the use of its officers, which affords them every information as to what, and how, they should set about observing; but, as far as I had an opportunity of seeing, it meets with general neglect from those for whom it was written. Whether the time will come when the fleet of Her Majesty's vessels which visit every sea, and which are stationed for months or years together upon interesting coasts otherwise little known, shall be looked upon as sources of scientific material which shall bear any proportion to the opportunities enjoyed and the treasures spent upon them, is a question which cannot now be solved; we can only hope that a very desirable change in this respect may by degrees be brought about, which cannot fail to be greatly to the advantage of naval officers as a class.

APPENDIX.

VOCABULARY of words used by the natives of Sau-o Bay, East Coast of Formosa (see Chapter VII.).

I. *Numerals.*

One	ētah (or issah).	Six	in-um.
Two	lu-sah.	Seven	pē-tou.
Three	too-roo.	Eight	ah-roo (or ah-loo).
Four	sóo-pah.	Nine	sē-wah.
Five	lē-mah (or e-mah).	Ten	stĕ-rei.

Ten is also represented by wón-ei, thus—

Eleven	wón-ei is-sah.	Thirteen	won-ei too-roo.
Twelve	won-ei lu-sah.	Nineteen	won-ei se-wah.

But *twenty* requires the other word for *ten*, as:—

Twenty	lu-sah stĕ-rēi.
Twenty-one	lu-sah stĕ-rēi is-sah.
Twenty-two	lu-sah stĕ-rēi lu-sah.
Twenty-nine	lu-sah stĕ-rēi se-wah.
Thirty	too-roo stĕ-rēi.
Forty	soo-pah stĕ-rēi;

and so on to *ninety*.

Hundred	see-voo.
One hundred (100)	is-sah see-voo.
Two hundred (200)	lu-sah see-voo.
Three hundred (300)	too-roo see-voo.
One thousand (1000)	is-sah r̃a-r̃a-r̃an.
1866	issah r̃a-r̃a-r̃an ah-loo see-voo in-um stĕ-r̃ei in-um.

2. Parts of the Body.

Head	hoo-roo.	Calf of the leg		rah-pan.
Hand	roo-kahp.	Buttock		poo-noon.
Eye	mah-tah.	Breast		ta-roo-nah.
Nose	hoo-nóong.	Arm		ree-mah.
Teeth	bun-ga-rów.	Knee		too-sol.
Hair	woo-kōose.	Thigh		pa-na-ni-yan.
Whiskers (also beard)	moo-mōose.	Foot		ree-kan.

3. Personal.

Man (male)	ma-roo-nah-nee.
Woman (female)	ta-roo-ang.
(The first includes *boys*, the second *girls*.)	
Children (boys or girls)	soo-niss.
Old man	na-ka-lan.
Old woman	vai-va-lam.
Mother	te-na.
Themselves (*i.e.* the tame aborigines)	Ka-ba-lan.
The wild aborigines of the hills	Ma-too-mal.
The Chinese	Bo-soos.

4. Articles of Dress and Furniture.

Coat	hoo-lōose.	Wood	broo-oor.
Stick	baa-ram.	Silk	see-reet.
Belt (either that around waist, or fillet around the head)	bar-oon.	Fishing-net	tchú-e.
		Roof of house	rah-poo.
		Linen	see-u.
Hat (native bamboo)	ro-co.	Pearl button	'tow-ear.
European hat	koo-boo.	Merino	nee.
Jacket	hoo-loose.	Red tape	trang-e-tang.
Trowsers	kwun.	Knife	sa-rick.
Shoes	la-po.	Worsted	ong-lee-pee.
Petticoat	lap-pi-yan, or ma-san.	Cotton	see-rah.
		Thread (of native manufacture)	tim-re-an.
Bag or pocket	roo-boose.		
Cigar	rai-poot.	Native cloth	ha-bah.
Beads	e-toose.	Shirt	rap-pou.
Bottle	bras-co.		

5. Names of Animals.

Dog	. . .	wah-soo.
Pig	. . .	ma-wo-nee.
Goat	. . .	koo-loo boo-lau.
Padi bird (Heron)		ah-larm.
Cock	. . .	drach-hook (guttural).
Hen	. . .	tee-na-na.
Hen's eggs	. .	soo-soo-se-na.
Fish	. . .	vow-hoot.
Fishing	. .	ta-pong-i-tchue.
Butterfly	. .	boo-row.
Crab	. . .	wah-rang.
Echinus	. .	ka-na-sow.

6. Articles of Food.

Sugar (loaf)	. . . .	wan-ing.
Rice	. . . .	brass.
Boiled rice	. . . .	mai-ee.

7. The Elements, &c.

Rain	. . .	oo-rahn.
Wind	. . .	var-lee.
Sun	. . .	ner-lun.
Surf (on beach)	. .	nar-een.
Fire	. . .	ra-mah.
Brass	. . .	pa-oo.
Iron	. . .	bah-liss.
Silver (coin)	. .	pe-lah.

8. Miscellaneous.

Reed-pipe	. .	wah-koo-par-in.
Wood-pipe	. .	kwa-ko.
Broom	. .	kai-sing.
Matchlock	. .	rah-pil-sa.
Tree	. .	bar-in.
Slate	. .	va-vow.
Hills	. .	ta-kerr.
Junk	. .	wa-pi.
Copper cash	. .	ka-ri-sow.
Rice-straw (thatch)		ra-mi.
Lily	. .	soo-a-yee.
Grass (or dried lily straw)	. .	brun.
Yes	. . .	ai-e.
No	. . .	mo.

9. Names of Men.

Sai-ah-nee.	Too-bah.
Mah-now.	Pah-keek.

10. Names of Women.

E-pai-ee.	Sing-ow.
Kin-lce-yan.	Ar-pee (a young girl).
Moo-hoot.	Sow-bahn.

Note.—All the above words are to be pronounced just as they are written, with English pronunciation. An accent has been placed upon some syllables, just to show the prominent part of the word ; and the circumflex over the r's shows that they should be trilled or rolled. The r's and l's were in many cases used indifferently by the same person, and appear to be interchangeable.

One thing is especially worthy of remark as differentiating these people from the Chinese, viz. the number and strong pronunciation of the r's in their language, while it is well known that the Chinese cannot sound that letter at all—thus, for *rice*, all Chinese say *lice*.

I may perhaps add to this a few words of the dialect used by the Chinese inhabitants of Ke-lung ; as follows—

Numerals.

One	tchee.	Nine		chew.
Two	nung.	Ten		chap.
Three	sah.	Eleven		eet.
Four	see.	Twelve		eet-gee.
Five	gaw.	Thirteen		eet-sah.
Six	lak.	Fourteen		eet-see, &c.
Seven	tcheet.	Thirty		sah-chap.
Eight	pooie.	Forty		see-chap.

Parts of Body.

Head	kow-moon.	Hair		tow-en.
Breasts	sä-woie.	Teeth		tchu-e-kee.
Nipple	knee.	Chin		tsui-tu.
Eye	mat-chew.	Tongue		tgee.
Ear	ā-yah.	Neck		am-koon.
Nose	pee.	Buttocks		kart-chung.
Mouth	tchu-e.	Hand		tching-towa.

Miscellaneous Words.

Shell	.	soo-ma.
Button	.	du-wu.
Hat	.	bo.
Bamboo hat	.	looi.
Leaf	.	tchĕung.
To write	.	seow-pah.
Stick	.	kwoi-a.
Pencil	.	pee-at.
Rain	.	haw.
Dog	.	gow.
Chicken	.	koo-ey.
Boy (or girl)	.	ginna.
Boy	.	tchapoi-ginna.
Girl	.	sawo-ginna.
Man	.	tchapoi-tworang.
Woman	.	sawo-tworang.

INDEX.

Aborigines (tame) of Formosa, 104
        costume of, 106
        characteristics of, 108
    (wild) search for, 107
Acridotheres cristatellus, 319
Actinia, new genus of, 198, 259
    floating, 353
    gigantic fish-sheltering, 150
Aden, 8
    marine animals at, 9
Admiral, visit of Chinese, 116
Advantages enjoyed by naval officers, 432
Agincourt island, 123
Agriculture of the Dyaks, 214
Agri-Horticultural gardens, 255
Albatross (yellow-billed), 412
Albicores, 366, 378
Alexandria, 4
Alima hyalina, 56
Alligator, 193
    bird, 230
Alpheus (habits of), 136
Amphibious habits of Malay children, 245
Ants, 232, 291
Apes-hill, Ta-kau, 39
Aplysia, 98
Ascension, 420
    vegetation and insects of, 423
Asterias, 191
Asterina, 197, 418
Atlanta Peronii, 55
Atmospheric phenomena, 146
Attaps, 214
Aurelia, 360
    not luminous, 399, 406
Azores, 428
    volcanic disturbance at, 429

Bamboo, 63, 214
Banca Straits, 213

Barking lizard, 169
Batangs, 236
Bear rock, 3
Beccari, Signor, 216
Beetles of Formosa, 71
    Labuan, 177
Beggars at Canton, 217
Bells at Manilla, 296
Berlidah, 235
Beröoids in the Atlantic, 363, 410
Birds of the Delta, 6
    between Pratas and Formosa, 28
    of Ta-kau, 44
    of Makung, 50
    of North Formosa, 79
    of Labuan, 167
    at sea, 412
Bird-keeping in China, 319
Birds' nests, edible, 230
Black islet, Haitan Straits, 131
Boat life in Canton river, 333
    songs of Malays, 233
    women in China, 18
        their infants, 19
Bogue forts, 331
Bombok, 239
Bornella digitata, 9
Botanic garden, Cape Town, 417
Boulder clay at Tam-suy, 61
Boys of Formosa, 45, 79
    Makung, 47
British character of Azores flora, 430
Brooke, Sir James, 163, 206
    Charles Johnson, 206
Bugis, 250
    prahus and imports, 250
Butterflies of Labuan, 181
    Daat, 185
    at sea, 143

Cairo, 6
Calamaris annulata, 150

Calamary, 49
Calm at sea, 10
Calzada at Manilla, 301
Camphor tree, 156
    trade in Formosa, 66
    laurel, 157
Canaries, 321
Candle-flies, 255
Canton, strangeness of, 330
    and Hong Kong, intercourse between, 328
Cape Town, 416
    vegetation, 414
Caricature plant, 15
Carpenter bee, 190
Case of moon-blindness, 309
Catamarans, 38
Cavern at Kelung, 86
    Sarawak, 233
Centipedes, 174
Centralisation of Chinese, 250
Cerapus tubularis, 56
Ceylon, 10
Chains of Salpæ, 371
Chama, 147, 149
Chameleons, 171
Chersydrus granulatus, 305
Chick-chack, 168
    reputed luminosity of, 169
Chinese new year, 311
    character of the, 326
    at Singapore, 249
Chironectes, 367
Christianity, prospects of, in China, 351
Chock-e-day, Formosa, 101
Chops in Canton Streets, 334
Chromodoris, 125, 149
Chuy-teng-cha, 78
Cicadas, 175
Cinnamon, 267
Citadel of Manilla, 294
Climate of Singapore, 258, 260
    Labuan, 198
Cloth (native) of Formosa, 106
Coal-mines at Kelung, 91
    quality of, 95
    Labuan, 157
    company, 159
Cobra, 171
Coccinella in the desert, 7
Cock-fighting at Manilla, 301
Cocoa-nut trees, 188
    planting in Singapore, 272
    beetles, 272
Cœnobitæ, 186
Coffee-planting, 267

Coleoptera of North Formosa, 71
    Labuan, 177
Colour of oceanic animals, 358, 367
    the sea, 387
Comatula, 137, 194, 258
Compound Salpæ, 368
    second form, 369
Contractility produces light, 408
Coral fish, 96, 147
    reef, 146
Corals, living, 147
Cormorants, 414
Correlation of forces, 407
Cotton cultivation, 266
Council of diamond-washers, 227
Coutts, Miss, 209
Craig island, 118
    geology of, 121
Crackers in China, 311
Cranes in Mediterranean, 3
Creseis, 99, 359
Cricket in Kelung cave, 87
Cycloclypeus, 126
Cymbulia, 99

Daat island, 185, 188, 196
D'Almeida, Mr. José, 264
Dammar trees, 157
    resin in coal, 160
Datu of Sarawak, 224
Defects of Chinese government, 340
Deformities of Chinese, 337
Delta of the Nile, 5
Dendractinia, 198
Density of population in Formosa, 81
Dentist, mountebank, 282
Desecration of graves in Labuan, 178
Desert of Suez, 7
Diamonds, 224
    washing, 225
    bird, 230
Diet of the Chinese, 338
Difficulties at sea, 152
Discontent of Chinese people, 343
Dogs at Makung, 48
    Dyak, 217
    eating in China, 339
Domestic animals of Formosa, 82
Doria, Marquis, 216
Doridopsis rubra, 259
Doris Barnardii, 51
    mantle cutting, 196
    exanthemata, 218
Draco volans, 212
Dragon flies, 188, 231
Dredge in Pacific, 125

INDEX. 441

Dredge lost, 127
Drongo, black, 79
Duck boats, 76
Durian, 271
Dutch occupation of Formosa, 36
    end of, 58
    fort at Makung, 47
Dyaks, 206
    first view of, 226
    girls, 226, 230, 238, 240

Earthquake at Manilla, 297
Echinus, variable species of, 418
Eclipse of the moon, 221
Electric snakes, 173, note
Elephants in Borneo, 216
Elevation of beach at Kelung, 90
Enoc, 198
Entomostraca, red, 130
    luminous, 394, 405
Eucharis, 363
Eulima on star-fishes, 191
European influence in China, 344

False Bay, 413
    luminosity of, 396
    fish numerous in, 414
Fa-tee nurseries, Canton, 335
Fayal, 428
Felis macrocelis, 217
Fiery Cross Reef, 146
Fire-flies, 235
    intermittent light of, 254
  making, 228
Firola, 55
Fishes living within Actiniæ, 151, 197
    among threads of Physalia, 365
Fishing operations at Kelung, 88
Flies at sea, 142, 320, note
Flint implements at the Cape, 416
Flora of Middle island, Haitan, 132
Flowers of Sarawak, 215
Fly on the ocean, 357
Flying-fish, 3, 11, 31, 83
    range of, 373
    vibration of wings of, 377
    abundance of, 379
Flying lizard, 212
    squirrel, 210
    foxes, 212
" Formby," the, 133
Formosa, character of aborigines of, 35
    treaty ports in, 37
    mountains of, 37

Fort Zeelandia, 56
Fowls, dyed pink, 217
Frigate bird, 29
Frogs eaten in China, 338
Fungus, luminous, 199

Galeopithecus, 210
Gambier at Sarawak, 210
    Singapore, 254, 270
Gambling in China, 279
    licensed, 282
    floating houses at Canton, 333
Gamboge, 269
Gannets, 13, 29
    of Pratas island, 30
Gelasimi, 40
Glaucus, 55
Glow-worm, luminosity of, 407
Gold in Formosa, 96
    at Sarawak, 225
    fish at Canton, 336
Golden lilies, 314
Gorgonia, stinging, 197
Grackle, 321
Grammatophyllum, 256
Grapsi, 122
Grass-cloaks, 19
Green Mountain, Ascension, 420
Gutta-percha, 268

Hadji, chief of Sarawak, 224
Haitan island, 129
Halobates, 358, note
Hantus of the Dyaks, 207
Head-hunting, 179
    house, Serambo, 237
    Bombok, 240
Helix Brookei, 197
Hemiptera of Labuan, 177
Hermit crabs, 186
Hirundo esculenta, 229
Hoa-pin-san, 124
Holothuriæ, 149, 192, 197
Hong Kong, 16
    beauty of scenery, 17
    highway robbery in, 322
Hornbills, 231
    drumming of, 239
Horsburgh lighthouse, 287
Horses in Formosa, 67
Hyalæa tridentata, 56
Hymenia recurvalis, 423, note
Hymenoptera of Labuan, 190

Iguana, 170
Image Point, Ke-lung, 90

Insects of Ascension, 423
Insecurity of Labuan, 180
  Hong Kong, 325
Insurrectionary movements in China, 342
Intolerance of Manilla government, 304

James Town, 419
Java sparrow, 15
Johore, shores of, 288
Joss house at Pratas island, 26
Jungle of Labuan, 155
Junks, 16
  masts, 139

Kabalan village of Sau-o, 103
Kaleewan river, 102
Kalong, 212
Ke-lung, arrival at, 82
  island, 83, 118
  people of, 83
Keramidia, 259
Klings, 245
  women, 246
  dhobies, 247
  religious ceremonies, 248
  bird-catchers, 257
Kok-si-kon. 58
Kok-singa, 58
"Koong-haye," 313
Ko-tou, 313
Kubong, 210
Kwang-yin hills, 65

Labuan coal mines, 157
  quality of, 158
  settlement of, 163
Ladies of Formosa, 45
Lake habitations, 14, 244
Land crabs, 40, 42
  Dyaks, 207
Language of China, a barrier, 349
Lanterns (Chinese), 316
Leaping fish, 41, 287
Leiothrix luteus, 320
Lichens of Ascension, 423
Light correlative with nerve force, 407
Lightning in the tropics, 12
Ligia, 51
Literary examinations in China, 341
Longwood, 419
Love grass, 155
Lucernaridæ, 360
Lucky stones, 313
Luminosity of the sea, 391
  categories of, 393

Luminosity of Singapore harbour, 395
  spontaneous, 402
  not dependent on wind, 404
Luminous sheath to ship, 395, 398, 404
  patches, 398
  flashes, 401
  fish, 406
  mucus, 408
  what classes of animals are, 409

Mandarin of Makung, 51
  Mbang-ka, 67
    visits ship, 68
  of Tam-suy, 69
  processions at Canton, 317
"Mandarin's leg," 138
Makung, 47
  absence of trees at, 50
Malays of Singapore, 242
  women, 243
  villages, 243
Manilla, 293
Marattia, 232
Mariveles, 293, 306
Martin, Mr., Sarawak, 209
Marundum reef, 218
Mbang-ka, Formosa, 66
"Meerschaum," 148
Megapode, 168
Melia, new species, 150
Memorial on Western Education, 346
Mestizas of Manilla, 295
Middle island, Haitan, 130
Milky sea, 403
Milvus govinda, 40
Mimosa, 259
Min, river, 138
Mina bird, 321
Modulus eaten at Makung, 49
Mollusca of Labuan, 193
Monkeys in Formosa, 40
  Labuan, 165
Moon blindness, 308
  shaped patches of light, 398
    explanation of, 400
Mosque of the Klings, 248

Napoleon Bonaparte, 419
Nepenthes, 155
Neptunus pelagicus, 367
Nibong palm, 202, 214
Nipa palm, 259
Noctilucæ, 259
  at Singapore, 395

INDEX. 443

Noctilucæ at False Bay, 396
    description of, 397
    flashing of, 403
    contractility of, 406
Nocturnal marine animals, 356
Notonectæ, 59
Nudibranchs, 9, 51, 98, 125, 135, 149, 195, 259
Nutmeg, 261
    tree, 262
    disease, 263
        causes of, 264
    recovery of trees, 265

Ocypoda, 25
Oliva erythrostoma, 50
    black variety of, 193
Ophiocoma, 219
Opium smoking, 283
    value and imports of, 284
    Chinese, 285
Opuntias, 419
Orang or Mias, 166
Orbitolites, 20, 126
Oryctes of cocoa-nut, 272
Oscillatoria, oceanic, 384
Osprey (H.M.S.) lost, 411
Ostræa canadensis, 78

Padi bird, 44
    fields at Ta-kau, 42
Padina, 51, 198, 427
Pagoda, 140
    anchorage, 139
Pako, 319
Palm island, 90
Paludinæ, 42
Pappan island, 185, 196
Parang (Dyak), 226
Pasig river, 302
Patella oculus, 417
Pawnbrokers' warehouses, Canton, 335
"Pearl" river, 316, 332
Pekin memorial, 346
    school of languages, 346
Pelagia, projection of threads by, 362
Pelagic animals, 351
Penang, 14, 261
Peninjau, Mount, 235
        view from summit, 238
Pepper, 271
Peronia, 97
Pescadores islands, 46
Petrels, fork-tailed, 412
Petroleum in Formosa, 96
    Labuan, 160

Phasma, 161
Phosphorescence, 392
Photography, 52
Phyllidia, new species of, 219
Phyllosoma, 56
Physalia, 357—363
    stinging powers of, 364
    fishes in threads of, 365
    not luminous, 406
Pico, Azores, 428
Pidgin English, 21, 350
Pigs in China, 43, 217, 338
    Labuan, 165
Pi-hi-kun, Formosa, 44
Pill-making crab, 288
Pinnacle island, 118
Pipe gamboge, 270
Piracy on Chinese coast, 132
    in Borneo, 205
Pistia stratiotes, 302
Planaria, 218
Pneumodermia, 99
"Poh," game of, 280
Ponghou harbour, 46
Population of Manilla, 295
Porpita, 357
Prahus of Sarawak river, 221
Pratas reef, 22
    island, 23
    flora, 24
    insects, 24
    shells, 25
    birds, 28
    seaweeds, 25
Proboscis monkey, 166
Provisions, cheapness of, at Makung, 49
Pterosoma, 54
Pulo Brani, 244
Puntinqua's garden, 335
Pyrosoma, 401
Python, 172

Queen's birthday, 69

Rafflesia, 216
Railways in China, 348
Rain in Labuan, 198
    Sarawak, 215
    Arabia,
    Ascension, 422
Rainbow, horizontal, 385
Raleigh rock, 117
Rapids on Tam-suy river, 80
    Sarawak river, 226, 232
Rats eaten in China, 339
Recruit island, 128
Red beetle of cocoa-nut, 272

Red discoloration of sea, 129, 354
    Sea, 7
    worms, 129
Retrospect, 431
Rhizostoma, 360, 362
Rice embargo, 64
    paper plant, 62
"Ruby," case of the, 133
Ruin-rock, 89
Rumbling fish, 134

Safety of Canton streets, 326, 329
Sagartia, new species, 131
Sagitta, 55
Sago planting, 273
Salpa pinnata, 368
Salt monopoly, 339
Samarang rocks, 203
Sampans at Hong Kong, 18
Sandstone of N. Formosa, 85, 92
Sanitarium, Rajah Brooke's, 238
Sarawak, 201
    river, 202
    flag, 202
    exports, 215
Sargasso sea, 366
Sarong, 203, 243
Sau-o bay, 101
Saw-mills of Johore, 286
School at Makung, 48
Scissor-grinder, 176
Scorpions, 27, 173
"Scylla" (H.M.S.), 410
Scyllæa pelagica, 366
Sea, varying aspects of, 386
    colour of, 387
    sawdust, 380
Seaweeds at Pratas, 25
Secchi's observations, 390
Sedans in Formosa, 66
Sensitive plant, 256
Sepia, 9
Serambo, visit to, 236
"Serpent," joined H.M.S., 22
Shanghai, 144
Shantung lark, 319
Sharks, 259
Shoals of Acalephs, 361
Shwingan passage, 141
Sik-kow, 78
Simon's Bay, 413
    rock at, 417
Simple Salpæ, 369
Singapore, 14, 242
    scenes in, 252
Sing-songs, 277
Slut island, Haitan, 135
Small feet of Chinese ladies, 21, 45

Snakes of Labuan, 172
Soap-stone rock, 139
Soil of Singapore, 261
Somali, 8
Southern Cross, 306
Sparks of light in the sea, 393
Sphærapœia Collingwoodii, 290
Sphex, cell-building, 190
Spiders of Pratas, 24
    Daat, 189
Stars, 6, 146
Star-fish, 191
Steep island, 102
Stephanomia, 99, 359
St. Helena, 418
Stinging hemiptera, 178
Storm at Sarawak, 235
    Hong Kong, 410
Streets of Canton, 334
Strombus eaten in China, 26
Sugar-cane, 268
Sulphur springs, 70
    geology of, 72
    present condition, 73
    trade in, 75
Sultan of Borneo, 162
Sumpitan, 257
Surgery in China, 337

Tablet island, 46
Tagalan, 296
Ta-kau, 38
Tam-suy, 60
    people of, 64
    river, journey on, 75
    night on, 77, 79
Tanjong Kubong, 211
    Putri, 275
Tarnuh-puti, 203, 209
Theatricals in China, 279
Tia-usu, 124
Tidal line of R. Min, 138
Tigers in Singapore, 253
Timidity of women and children, 48
Tobacco manufacture, Manilla, 303
Tombs at Tam-suy, 64
Towing-net, 355
Traveller's tree, 14
Treaty of 1858, 345
Trepang, 150
Trial by jury at Hong Kong, 327
Trichodesmium, 380
    abundance of in China Sea, 382
    microscopic examination of, 383
Tropic birds, 13, 412
Tropical nights, 306

Tryxalis, 40
Tuan Muda of Sarawak, 206
Tumblers, 320
Tumonggong of Johore, 275, 277
Turtle ponds, 426
Typhoon, 33

Unity of the Chinese nation, 340

Valonia (seaweed), 136
Velella, 357
Victoria peak, 16
Villa de Horta, 428
Visits of ceremony in China, 315
Vocabulary of Formosa dialect, 113 (Appendix)
Volcanoes near Manilla, 298

Wages of Chinese and Malays, 273
Wariness of crabs, 187
Water buffaloes, 43

Water beetles, 187
  snakes, 304
Waterspout, 153
Weather at Manilla, 305
Western trade with China, 345
Whales' bones at False Bay, 416
Whampoa, 316, 331
Whampoa's garden, 251
Wideawakes of Craig island, 119
  Ascension, 424
Wild night, 142
Window-oyster, 294
Women of Manilla, 295
Wosung river, 143
Wou-wou, 235
Wrecking at Makung, 52

Yang-tze-Kiang, 143

Zoea, 32
Zostera, 25

THE END.

www.ingramcontent.com/pod-product-compliance
Lightning Source LLC
Chambersburg PA
CBHW022115300426
44117CB00007B/719